Conservation Agriculture Technologies

The Editors

Dr. Rahul Kumar is presently working as a Senior Research Fellow, ICAR Research Complex for Eastern Region, Patna (Bihar), India. He is presently working on Conservation agriculture and wheat- rice cropping system in CSISA Project. He has been completed Ph.D. in Plant Pathology from Dr. Rajendra Prasad Central Agricultural University, Pusa, Samastipur, Bihar, M.Sc. in Plant Pathology from SHIATS, Allahabad. A number of national and international reputed publications are credited to him. He has two book, two book chapters, 22 Hindi articles and 15 abstract to his credit. He also honoured with Young Professional award 2015 by Society for Upliftment of Rural entrepreneurship, Varanasi for his outstanding contribution in the field of Plant Pathology. Becides it, he honoured with Young Scientist award 2017 by BIOVED Society Allahabad for his contribution in the field of Plant Pathology and Excellent in Research Award-2018 by Society of Agriculture Innovation and Development (SAID) Ranchi for his contribution in the field of Plant Pathology. Best Ph. D. thesis award received on the occasion of 2nd International Conference on Food & Agriculture, Dhanbad, Jharkhand, organized by Endling Conference. Several oral and poster presentations in also done in various national and international seminar/symposium/conferences in India.

Abhinandan Singh is pursuing Ph.D. at RPCAU, Pusa-Bihar. He graduated B.Sc. (Ag) from Udai Pratap Autonomous College, Varanasi, UP and obtained M.Sc. (Ag) in Agronomy from CAU, Imphal- Manipur. He has qualified ICAR JRF, SRF and ASRB NET in Agronomy. He has more than twenty research papers in national and international journals of repute. He has two books, eight book chapters and five abstract to his credit. He has participated, poster and oral presented in national and international symposium & seminar. He also honoured Young Scientist award from Society of Agriculture Innovation and Development (SAID) Ranchi for his contribution in the field of Agronomy. He received Bangali Baboo medal 2018 by BRIATS, Allahabad on the occasion of 20th Indian Agricultural Scientists and Farmers congress for his outstanding contribution in field of Agronomy.

Dr. Pankaj Kumar Ojha did his M.Sc. (Ag.) in Extension Education from Banaras Hindu University, Varanasi and Ph.D. in Extension Education from Dr. Rajendra Prasad Central Agricultural University, Pusa. Besides these, he did Post-graduate Diploma in Rural Development from IGNOU and Post-graduate Diploma in Theatre Communication from Banaras Hindu University, Varanasi. He has numbers of publications to his credit including books (two), research papers, book chapters, review articles, popular articles, etc. He has to credit a national award AIASA Gold medal for his literary contribution. Presently he is working as an Assistant Professor, Department of Agricultural Extension, Banda University of Agriculture and Technology, Banda, U.P.

Conservation Agriculture Technologies

– Editors –

Rahul Kumar

Abhinandan Singh

Pankaj Kumar Ojha

BIOTECH

2026

BIOTECH BOOKS®

ISBN 978-81-7622-448-2

Published by: **BIOTECH BOOKS®**
4762-63/23, Ansari Road, Darya Ganj,
New Delhi - 110 002
Phone: +91-011-43003222
E-mail: biotechbooks@yahoo.co.in

Typeset at: **Classic Computer Services**
Delhi - 110 035

Printed at: **Replika Press Pvt. Ltd.**

PRINTED IN INDIA

Preface

India is an agricultural country. Contribution of agriculture sector in the GDP is about 21 per cent. It provides employment to 45 per cent of country's labor force and is source of livelihood for 60 per cent of the rural population. It has a vital role in ensuring food security, generating overall economic growth, and reducing poverty. Our population is increasing very quickly; there is lot of population pressure on the agriculture sector. To feed this high population we are trying to enhance the agriculture productivity on the expense of land, water, labor, capital, climate and other resources ignoring the recommendations for good agricultural practices. Industrialization and urbanization has further aggravated the problem by reducing the area of production and polluting the land, water and environment which is a direct threat to our agricultural productivity. With the unbalanced use of our resources, we have created many problems such as loss of fertile land, water logging, soil salinity, erosion, pollution of above ground and underground water, habitat destruction etc. We are wasting our water resources which are decreasing rapidly. 75 per cent area of India is dependent on irrigation water. Our mismanagement of resources is a permanent cause of the higher levels of CO_2 emissions and temperature increase leading to climate change with extreme events which are destructive to our resources and agriculture productivity, which may cause the food security issues to rise up. Food security is a global problem and especially for India, it is a great challenge. About 30 per cent of our population is living below poverty line, and our farmer is also very poor with small land holdings. The high prices of inputs (fuel, seed, fertilizers, pesticides, herbicides, machinery and electricity etc) have added much to the anxiety of the farmers. Farmers are living a subsistent life. Our average crop yields are much lower than other countries despite having lot of potential. Despite of great recent progress, hunger and poverty remain widespread and agriculturally driven environmental damage is widely prevalent. The idea of agricultural sustainability centers on the need to develop technologies and practices that do not have adverse effects on environmental goods and services, and that lead to improvements in productivity per unit area and profitability. Resource Conserving Technology (RCT) is a broad term that refers to any management approach or technology that increases factor productivity including land, labour, capital and inputs. Some of these technologies are described in this book.

The book includes more than 20 conceptual and comprehensive chapters, contributed by eminent researchers in their respective fields.

We hope that the book will be of immense value for the students, scholars, academicians, extension workers, policy makers and development practitioners. We request readers to give their reactions and suggestions soon.

Authors' contributions have been presented in their original form with some necessary editorial modifications. We express our deep sense of gratitude to all the learned professionals for making their valuable contributions.

Rahul Kumar

Abhinandan Singh

Pankaj Kumar Ojha

Contents

Conservation Agriculture Technologies *Pages* **1–13**
Editor: **Rahul Kumar, Abhinandan Singh and Pankaj Kumar Ojha**
Published by: **BIOTECH BOOKS, NEW DELHI**

1 Conservation Agriculture: Importance from Indian Perspective

Abhinandan Singh[1]** *and Aditya Kumar*[2]

[1]*Department of Agronomy, Dr. Rajendra Prasad Central Agricultural University, PUSA, Samastipur – 848 125, Bihar*
[2]*Department of Agricultural Chemistry and Soil Science, Udai Pratapa Autonomous College, Varanasi – 221 002, Uttar Pradesh*

Traditional agriculture, based on tillage and being highly mechanized, has been accused of being responsible for soil erosion problems, surface and undergroundwater pollution, and more water consumption in India. Conservation agriculture promotes most soils to have a richer bioactivity and biodiversity, a better structure and cohesion, and a very high natural physical protection against weather (raindrops, wind, dry or wet periods). Soil erosion is therefore highly reduced, soil agronomic inputs transport slightly reduced, while pesticide bio-degradation is enhanced. It protects surface and groundwater resources from pollution and also mitigates negative climate effects. Hence, Conservation agriculture provides excellent soil fertility and also saves money, time and fossil-fuel. It is an efficient alternative to traditional agriculture, attenuating its drawbacks.

Introduction

In the last five decades, intensification of agriculture using high-yielding crop varieties, fertilizers, irrigation and pesticides has increased the agricultural production globally. During this process, the second generation problems have come up like groundwater decline, environmental pollution, fatigue in land and crop productivity, emergence of resistant insect, pests and weeds and poor soil health. These problems have threatened the very base of the agricultural production. To enhance the production for ever growing population corrective measures must be taken immediately. The urgent goal is to increase the production while conserving environmental resources on a sustainable basis. The recent unprecedented fuel and food crisis have forced global community to take appropriate measures to substantially enhance food production in a sustainable manner to feed the growing population. Conservation agriculture has three basic principles, (i) continuous no or minimal mechanical soil disturbance; (ii) crops either directly drilled or broadcast at the soil surface with straw or stubbles of the previous crop, permanent organic matter soil cover, especially by crop residues; and (iii) diversified production portfolio.

Resource Conservation Technologies

RCT is a broad term that refers to any management approach or technology that increases factor productivity including land, labour, capital and input. RCTs includes a wide range of practices including no till/minimum tillage, surface seeding, skip row furrow irrigation, inter cropping, water harvesting and supplemental irrigation, mulching and residue management, live fences and vegetable barriers *etc.*

Need of Conservation Agriculture

For sustainable intensification of crop production, CA is the base concept. The degrading land resources, declining quality irrigation water and increasing demand for food and feed has forced farmers for the adoption of CA. the need for CA can be discussed as under:

1. Land Degradation

Land degradation is a serious environmental problem that threatens the ecosystem, health and food security. More and more fertile land has comes under marginal and degraded lands (Vlek, 2008). The main land degradation processes like soil erosion, nutrient mining, carbon loss *etc.* are aggravated by human interference and intensified agricultural operaions.

2. Soil Erosion

The soils today are being degraded at a much faster rate than they can be formed by natural processes. Excessive tillage as practiced in conventional tillage is one of the most important drivers of soil erosion. Due to erosion during last 40 years, about 30 per cent of the world's arable land has become unproductive and most of it has been abandoned for agriculture (Kendall and Pimental, 1994).

3. Soil Nutrient Mining

Soil nutrient mining occurs when extraction of useful nutrients from the soil by agriculture exceeds the rate of replenishment in the system. Nutrient depletion in soil adversely affects the soil quality and reduces crop yield. Because of this reason the per kg grain output due to unit kg NPK supply in India has reduced from 13.4 in 1970 to 3.7 in 2010. If we want to keep soil enriched with nutrients for the future generations to come, we need to adopt CA as a solution.

4. Declining Fertility and Productivity

Due to continuous traffic on the soil, the physical and chemical properties of the soil get degraded. More dependence on inorganic fertilizers has resulted in poor organic carbon status of the soil. If crop residues are incorporated before sowing it supplies sufficient quantities of micro nutrients also.

5. Soil Carbon Loss

Soil carbon/organic matter is known as black gold because of its vital role in physical, chemical and biological processes with in the soil system (Reicsky and Saxton, 2007). Soil erosion results in the removal of organic matter and essential plant nutrients from the soil and the reduction of the soil depth. Conversion of virgin land for cultivation and grazing, repeated cultivation promoting soil respiration without additional carbon inputs and soil erosion are the main causes of loss of organic matter.

Conservation Agriculture: Indian Scenario

Resource degradation problems are manifesting in several ways. Declining water tables in many agriculturally important regions imply increasing pumping costs, replacement of shallow gravity tubewells with submersible pumps at huge cost, adverse effects on water quality and overall ecology of the region. Declining soil carbon and fertility are reflecting in declining soil bio-diversity, multiple nutrient deficiencies, need for increasing inputs use to maintain yields and implications for quality of produce and environment. Inefficient input. Excessive nutrient use and management practices are leading to widespread contamination of surface and groundwater with connected health hazards. Rainfall runoff and erosion of surface soil are responsible for continued low productivity, more so in rainfed areas. Conservation agriculture has emerged a major way forward from existing unsustainable mode of agriculture.

The role of conservation agriculture in developed and developing nations to achieve the ultimate goal of food security has been very well recognized. Globally, over 160 million hectares of land is under conservation agriculture.

In India, efforts to adapt and promote resource conservation technologies have been underway for nearly a decade but it is only in the past 4 to 5 years that the technologies are finding rapid acceptance by the farmers. Efforts to develop and spread conservation agriculture have been made through the combined efforts of several State Agricultural Universities, ICAR institutes and the CG promoted, Rice-Wheat Consortium for the Indo-Gangetic Plains. Unlike, in the rest of the world, in India spread of technologies is taking place in the irrigated regions in the Indo-Gangetic plains where rice-wheat cropping system dominates. CA systems have not been tried or promoted in other major agro-ecoregions like rainfed semi-arid tropics, the arid regions or the mountain agro-ecosystems.

Conservation Agriculture in Rainfed, Semi-arid and Arid Regions

Rainfed, semi-arid and arid regions are characterized by variable and unpredictable rainfall, structurally unstable soils and low overall productivity. These soils are poor in organic carbon and water holding capacity also. Thus, to improve these qualities reduced tillage, controlled traffic and crop residue incorporation are must. Results of most research station studies show that zero/reduced tillage system without crop residues left on the soil surface have no particular advantage because much of the rainfall is lost as runoff due to rapid surface sealing nature of soils. It would therefore appear that no tillage alone in the absence of soil cover is unlikely to become a favoured practice. However, overall productivity and residue availability being low and demand of limited residues for livestock feed being high also poses a major limitation for residue use as soil cover in the arid and semi-arid regions. In the semi-arid regions there is wide variability in rainfall and its distribution and nature of soils, there is need to identify situations where availability of even moderate amount of residues can be combined with reduced tillage to enhance soil quality and efficient use of rainwater. There appears no doubt that managing zero-tillage system requires a higher level of management *vis-a-vis* conventional crop production systems.

Conservation Agriculture in Irrigated Ecosystems

In India efforts to promote CA technologies have largely been focused in the Indo-Gangetic plains covering the states of Punjab, Haryana, Uttar Pradesh, Bihar and West Bengal. Over the decades of seventies and eighties rice-wheat has emerged as the major cropping system extending to over 11 million ha. Concerns of stagnating productivity, increasing production costs, declining resource quality, declining water table and increasing environmental problems are the major forcing factors to look for alternative technologies particularly in the north-west region encompassing the states of Punjab, Haryana and western UP. In the eastern region covering eastern UP, Bihar and West Bengal developing and promoting strategies to overcome the constraints responsible for continued low cropping system productivity have been the primary concern. It is in this context that the efforts aimed at developing and promoting CA technologies over the past decade or so must be viewed. The primary focus of developing and promoting CA practices has been development and adoption of zero-till seed cum fertilizer drill for sowing, raised bed planting system, laser equipment aided land levelling, residue management alternatives, alternative to rice-wheat cropping system in relation to CA technologies, *etc.* The area planted with wheat adopting zero-till drill has been rapidly increasing over the past decade in the whole Indo-gangetic plain. Also, this technology is not limited to farmers with a large holding but the small holding farmers are adopting the technology through custom hiring (Rs. 750 to 900 per ha for sowing wheat) while the farmers with bigger holding are purchasing the seed drill. In a sample study amongst farmers adopting zero-tillage it was observed that the area brought under zero-tillage visa- vis total area under wheat crop varied with an overall average of 45 per cent under zerotill system.

The wheat yields in zero-till fields are invariably higher by 3 to 5 per cent compared to planting in conventionally tilled fields as a result of timely planting of wheat when zero-till was used. Reduced cultivation costs up to Rs 1500 to 2250 per ha results from reduced costs of diesel, labour and use of herbicides. Reduced incidence of weeds, particularly *Phalaris minor* and savings in water use were other benefits frequently cited by participants. Farmers who have used laser land leveller reported positive effects in improving crop establishment, uniformity of crop maturity, increase in area available for cultivation (2 to 5 per cent), improved efficiency of water application and increased water productivity resulting in large savings in irrigation water (upto 35 per cent) and in improved use efficiency of applied nutrients. With increased availability of laser land levelling equipment, precision land levelling could become an important basal step for adopting CA systems. Planting wheat and other crops on the raised beds is being increasingly adopted by farmers. The advantage cited by farmers include savings on inputs like seeds and fertilizers, significant savings in water and reduced use of plant protection chemicals while the labour requirements increased somewhat. The overall crop yields and net return increased. Bed planting systems also offered greater opportunity for crop diversification and were amenable to key elements of CA, zero tillage and residues managed on surface. Some efforts are also underway to adapt CA practices for rice crop in rice-wheat cropping system. The approach being tested is to grow rice under aerobic condition on raised beds under non-puddled conditions. Although some success has been reported, managing weeds is a major challenge. Edaphic requirements of growing rice do not permit zero tillage or surface retention of crop residues. The impact of CA practices on resource quality is gradual and significant only over a period of time. When rice is grown after 'zero' till wheat these benefits are not likely to accrue. The question therefore is: Is rice grown as apart of rice-wheat cropping system in the north-west an anti thesis of CA? Thus, if principles and practices of cultivating rice in the semi-arid north west achieved mainly through intensive irrigation contradicts the technological and institutional innovations that enable or form the basis for CA and sustainability, then alternate cropping systems that meet these challenges have to be evolved. The issues which are important, therefore, include: What are the alternate cropping systems for adapting and promoting CA; how do they balance productivity per unit water, vulnerability to leaching losses of applied nutrients, *etc.* The choice of cropping systems can have a strong influence on resource quality through processes such as nutrient depletion and/or enrichment through underground biomass/crop residues, need for external inputs, impact on environment *etc.* It would appear from the progress of CA related activities that while zero-till planting of wheat is catching up rapidly the benefits are largely related to ability to achieve timely planting. The concept of CA emerging globally stipulates least disturbance of soil (*e.g.* through zero-tillage) together with maintaining soil cover (*e.g.* through crop residue) as fundamental to resource improvement.

Conservation Agriculture: Implications for Technology Generation, Evaluation and Upgradation

Conservation agriculture as an upcoming paradigm for raising crops will require an innovation system perspective to deal with diverse, flexible and context specific needs of technologies and their management. Innovation system perspective involves the groups of organizations and individuals involved in the generation, diffusion, adaptation, and use of knowledge of socio-economic significance, and the institutional context that governs the way these interactions and processes take place. Conservation agriculture R&D, thus will call for several innovative features to address the challenge. Some of these are discussed as follow.

Understanding the System

Conservation agriculture systems are much more complex than the conventional systems. Managing these systems efficiently will be highly demanding interms of understanding of basic processes and component interactions, which determine the whole system performance. For example, surface maintained crop residues act as mulch and therefore reduce soil water losses through evaporation and maintain a moderate soil temperature regime. However, at the same time crop residues offer an easily decomposable source of organic matter and could harbour undesirable pest populations or alter the system ecology in some other way. No-tillage system will influence depth of penetration and distribution of root system which, in turn, will influence water and nutrient uptake and mineral cycling. Thus the need is to recognise CA as a system, understand system interactions and develop management strategies, which take into account these interactions. terms of understanding of basic processes and component interactions, which determine the whole system performance. For example, surface maintained crop residues act as mulch and therefore reduce soil water losses through evaporation and maintain a moderate soil temperature regime. However, at the same time crop residues offer an easily decomposable source of organic matter and could harbour undesirable pest populations or alter the system ecology in some other way. No-tillage system will influence depth of penetration and distribution of root system which, in turn, will influence water and nutrient uptake and mineral cycling. Thus the need is to recognise CA as a system, understand system interactions and develop management strategies, which take into account these interactions. Building a Systems Perspective. Building a systems perspective in research will be fundamental in generating and promoting new technologies. A systems perspective is best built working in partnership with farmers who are at the core of farming system, with scientists, farmers, extension agents and other stakeholders working in partnership mode will therefore be critical in developing and promoting new technologies. Understanding system interactions and developing management strategies will call for teams of scientists across disciplines working together. This will also call for new ways of managing and leveraging funding mechanisms to permit working together of various stakeholders to facilitate information exchange and learning.

Partnerships and Linkages/Networks

In the conventional agricultural R&D, the system is to set research priorities and allocate resources within a framework, and no attention is given to build relationships and seek linkages with partners working in the complementary fields. But for successful CA R&D, building relationships with partners, seeking linkages and understanding the relationships and processes would widen the scope and improve the relevance of CA to development agendas. The process of networking, forming alliances and partnerships, negotiating priorities and approaches would help in further supporting and planning R&D for CA.

Technological Challenges

While the basic principles which form the foundation of CA practices, that is, no tillage and surface managed crop residues are well understood, adoption of these practices under varying farming situations is the key challenge. These challenges relate to development, standardisation and adoption of farm machinery for seeding with minimum soil disturbance, developing crop harvesting and management systems with residues maintained on soil surface and developing and continuously improving site-specific crop, soil and pest management strategies that will optimise benefits from the new system. The efficiency of CA systems can be further enhanced with appropriate spatial and temporal crop-agroforestry sequencing to benefit from ecological processes which contribute to system resilience and reduced vulnerability to productivity reducing factors.

Site Specificity

Adaptive strategies for CA systems will be highly site specific yet learning across the sites will be a powerful way in understanding why certain technologies or practices are effective in a set of situation and not effective in another set. This learning process will greatly accelerate building a knowledge base for sustainable resource management. The comparative analysis would also help to suggest a number of general/diverse principles of CA. Developing and promoting networking to share information amongst farmers, scientists and other stakeholders would be critical in advancing spread and continued upgradation of CA systems. However, understanding the diversity and the context-specific nature of the processes would help in learning and changing for better.

Conservation Agriculture Research Must have a Long-term Perspective

CA practices *e.g.* no tillage and surface maintained crop residues result in resource improvement only gradually and benefits come about only with time. Indeed, in many situations benefits in terms of yield enhancement may not come about in early years. Evaluating the impact of CA practices therefore must have a longer term perspectives.

Adoption and Impact Studies

Most of adoption and impact studies still concentrate on production based statistics. They rarely include their impact on natural resources, processes/ relationships that enable the success or failure of the study. These studies should be part of the larger study and not a separate study that is done after the adoption is over or the end of research process. The institutional learnings, that form part of the ongoing studies facilitate learning by doing and thus are important to be a part of the innovation process before technological innovations take place.

Adoption of CA systems over large areas is a way to improved ecological foundation, which form a basis for sustainable agriculture. It is, therefore, important that evaluation of CA takes into account the impact of the system on overall environmental performance or improvement in the natural resources. Thus, research and development for CA will call for several innovative features to address the challenges.

Recommendations for Conservation Agriculture Research and Development

Conservation agriculture offers an opportunity for arresting and reversing the downward spiral of resource degradation, diminishing factor productivity and decreasing cultivation costs making agriculture more resource use-efficient, competitive and sustainable. While R&D effort over the past decade have contributed to increasing farmer acceptance of zerotillage for wheat in rice-wheat cropping system, this has raised a number of institutional, technological and policy related questions related to technology generation, adaptation and further improvement, which must be addressed if CA practices have to be adopted on a sustained basis. The following recommendations and institutional arrangements represent an integrated view of focussed group discussions to define R&D priorities for CA. Myopic 'food security' policy based on cereal production must now replace a well articulated policy goal for livelihood security and rural development in India, and other developing countries. A factor price support followed by mandatory procurement for the crops alternative to rice and wheat that do not overexploit natural resources and enabling policy mechanism and environment for other crops was stressed upon. Emphasis should be on selecting and developing crops and varieties which result in saving of inputs like water, electricity, fertilizers, *etc.* Promote CA as a theme ensuring effective linkages between R&D. Leverage funding mechanisms to permit multidisciplinary teams of scientists and farmers to work in a participatory mode and promote networking amongst farmers, scientists, farm machinery manufacturers to facilitate information exchange and learning. Adaptive strategies for CA will be site specific and call for initiation of R&D efforts under a range of situations which aim at exploring multiple benefits by building context specific partnerships to address key livelihood issues. Learning process across the sites will be a powerful way of understanding which technologies and why they are effective in a particular set of situations. This will greatly accelerate building a knowledge base for sustainable resource management. Developing and promoting networking to share information among partners will be critical in advancing spread and continuous improvement of CA systems. As CA systems are much more complex managing the systems efficiently will be highly demanding in term of understanding of basic processes and component interactions which determine the whole system performance. Developing strategies to manage CA systems will require a clear understanding of total system functioning and building a systems perspective in generating and promoting new technologies. A systems perspective is best build working in partnerships with farmers who are at the core of the farming system and best understand this system. Scientists, farmers, extension agents and other stakeholders like policy makers, private sector, working in partnership mode will be critical in developing and promoting new technologies.

Promoting CA will call for moving away from the conventional compartmentalized and hierarchical arrangement of research that generates perfect technologies, extension that delivers it and farmers who passively adopt it. All the stakeholders involved would need to be brought together on a common platform to conceive end-to-end strategies. Institutionalising the role of research, extension and farmers in such a way that the partnership among these stakeholders might be strengthened right from the beginning of the project, and building up sense or enabling of ownership among them. Conservation agriculture offers opportunities for diversified cropping systems in different agro-ecoregions. Developing, improving and standardizing equipment for seeding, fertilizer placement and harvesting ensuring minimum soil disturbance in residue managemant for different edaphic conditions will be key to success of CA systems. For many situations, for example in hilly tracts, for small land holders bullock drawn equipment will have greater relevance. Ensuring quality and availability of equipment through appropriate incentives will be important. Impact of CA technologies on resource quality and environment is slow and significant changes may show up only over a period of time. There is a need to set up long term monitoring sites in representative agro-ecological situations to monitor resource quality, ecology and productivity with a view to continuously improve management options for sustained improvements. Evaluation and impact of CA practices therefore needs a longer term and broader perspective, which goes beyond yield increase studies.

Conservation agriculture practices *i.e.*, no tillage and surface maintained crop residues set in processes which initiate changes in soil physical, chemical and biological properties which, in turn, affect root growth and crop yield. Understanding the dynamics of these changes and interactions between physical, chemical and biological phases is basic to developing improved soil, water and nutrients management strategies. Similarly understanding the dynamics of qualitative and quantitative changes in soil biodiversity (flora and fauna), diseases and pest causing organisms including weeds in relation to altered management practices is fundamental to evolve control measures which involve minimum use of environmentally harmful chemicals. Conservation agriculture technologies bring about significant changes in the plant growing microenvironment. These include changes in moisture regimes, root environment, emergence of new pathogen populations and shifts in insect-pest scenario *etc.* The requirements of plant type suited to the new environment and to meet specific mechanization needs could be different. There is need to develop complementary crop improvement programme aimed at populations which are better suited to the new systems. Farmer-participation research approaches would appear promising for identifying and developing crop varieties suiting specific environments or locations. Accelerated development and adoption of CA technologies will call for greatly strengthened monitoring and evaluation along with policy research. Understanding constraints in adoption and putting in place appropriate incentive for adopting CA systems will be important. This will call for considerable strengthening of social science research. There is a need for generating good resource database with agencies involved complementing each others' work. Besides resources, systematic monitoring of the socio-economic, environmental and institutional changes should become an integral part of the major projects on CA. Inclusion of conservation and sustainability concepts in the course curricula with a suitable blend of biophysical and social sciences would be important. Institutional mechanisms are required to ensure that CA is seen as a concept beyond agriculture and promote it as a theme ensuring effective linkages between R&D activities.

Conservation agriculture must aim at broad livelihood strategies and move towards forming conservation villages, with appropriate agribusiness strategies to increase employment in areas where it is adapted. However, caution must be taken to avoid blanket adoption of practices just everywhere, it should be site and production system based. Conservation agriculture implies a radical change from traditional agriculture. There is a need for policy analyses to understand how conservation technologies integrate with other technologies, policy instruments and institutional arrangement that promote or deter conservation agriculture. It is therefore a challenge both for the scientific community and the farmers to overcome the past mindset and explore the opportunities that CA offers for sustainable agriculture. Spread of CA, therefore, will call for a greatly strengthened research and linked development efforts.

Role of Conservation Agriculture

The unsustainable exploitation of land resources is leading to widespread degradation which has serious implications for food security and ecological integrity. The reserves of potentially arable prime agricultural land are dwindling. Even the highly effective fertilizers and seed technologies of Green Revolution are reaching a point of diminishing returns. With soil organic matter, a key resource to conserve the agricultural research community has re-assessed the costs and benefits of conservative soil management systems. Food security, ecological integrity and environmental health may be achieved by reducing tillage, leaving residues on the field, augmenting nutrient inputs and reducing pest pressures through crop rotation, chiefly through CA.

A. Reduced Cost of Cultivation

CA in the form of no-tillage has now been adopted on more than 95 million hectares world wide. The impact of zero tillage in rice-wheat system of India showed that investment in zero tillage was highly beneficial with a B: C ratio of 39. The observed gains in crop production and environmental protection are encouraging which argument that CA should be practiced on a wider area. CA ensures farm level economic viability and stability. Reduction in cost of cultivation is a key factor contributing to rapid adoption of zero-till technology. Most studies show that the cost of wheat production has been reduced by Rs.1500 to 2000 per hectare. Cost reduction is attributed to savings on account of diesel, labour and input costs, particularly herbicides.

B. Improve Soil Organic Carbon

Surface retention of sufficient crop residues increases soil organic matter and biological activity which enhances long term sustainability. It saves non-renewable sources of energy and enhances carbon sequestration.

C. Reduce Environmental Pollution

Soil erosion and leaching of applied agricultural chemicals cause severe environmental pollution. Emission of CO_2 by agriculture can be decreased by reducing tillage and maintaining crop residues on the soil surface to increase C sequestration in the soil, especially when combined with the reduced burning of fossil fuels for field operations associated with reduced or zero tillage. Nitrous oxide emissions are more damaging to the environment than CO_2 which can be reduced by improving N use efficiency and cutting methane emissions by limiting the extent of flooded rice cultivation.

If adequate levels of crop residues are retained on the soil surface and combined with reduced or zero tillage, the C sequestered will correspond in reduction in CO_2 emissions. CA involving zero-till and surface managed crop residue systems are excellent opportunity to eliminate burning of crop residues which contribute to large amount of green house gases like CO_2, CO, NO_2, SO_2 and large amount of particulate matter. Burning of crop residues, also contributes to considerable loss of plant nutrients, which could be recycled when properly managed. Large scale burning of crop residues is also a serious health hazard

D. Enhances Partial and Total Factor Productivity

The increase in the prices of fertilizers has been substantial, especially in the developing countries. Hence, there is a need to enhance the output per unit of input applied in terms of fertilizers, herbicide and pesticides. Suitable fertilizer management practices that are compatible with appropriate CA based technologies may result in enhanced N-use-efficiency. CA also reduces pollution of surface and groundwater from chemicals and pesticides, resulting from improved inputs use efficiency. In properly managed zero-till planted wheat and mustard yields obtained are invariably higher by 4 to 6 per cent compared to traditionally prepared fields for comparable planting date.

E. Improves Soil Properties

Soil degradation from wind and water erosion, as well as, decline in soil, physical, biological and chemical properties are linked to the excessive levels of tillage and extensive removal/burning of crop residues. The mean weight of water stable aggregate is higher for permanent beds as compared to the conventional methods. The more residues that are left, the better the aggregate stability will be. Residue incorporation also improves the biological activity and diversity in the soil including natural predators and competitors.

F. Water Saving

Rational retention of adequate levels of crop residues protects the soil from water run-off and erosion, improves water infiltration and reduces evaporation to improve water productive. Under rainfed situations, by reducing run-off and evaporation losses, CA helps to generate "more crop per rainfall drop". Use of CA based raised bed, furrow irrigated seeding system in India has resulted in significant saving in irrigation for a wide spectrum of crops, over traditional farmer practice of seeding on the flat with flood irrigation. Raised bed seeding system provides striking increase in irrigation water-use-efficiency, especially permanent raised bed seeding where no tillage is used on the top of the bed. Average irrigation water saving of 20-25 per cent with equal or even higher crop yield has been observed with the raised bed system.

G. Optimum Resource Utilization

CA employs economically viable, diversified crop rotations to help moderate possible weeds, diseases and pest problems, enhances soil biodiversity, takes advantage of biological N fixation and soil enhancing properties of different crops, reduce labour peaks and provide new risk opportunities to the farmers. Limited experimental results and farmers experience indicate that considerable saving in water (up to 20-30 per cent) and nutrients are achieved with zero-till planting and particularly in laser levelled and bed planted crop.

H. Resource Impovement

No tillage when combined with surface managed crop residues sets in the processes whereby slow decomposition of residues results in soil structural improvement and increased recycling and availability of plant nutrients. Surface residues acting as mulch, moderate soil temperatures, reduce evaporation, improve biological activity and provide more favourable environment for root growth, the benefits which are traditionally sought from tillage operations.

I. Reduced Incidence of Weeds

The weeds produce a huge number of seeds. These seeds remain on the surface when the soil is not deep tilled. In case of Orobanche in mustard, most weed seeds are unable to reach a suitable depth which facilitates its germination. Most studies tend to indicate reduced incidence of *Phalaris minor*, a major weed in wheat, when zero-tillage is adopted resulting in reduced use of herbicides.

J. Crop Diversification Opportunities

Adoption of CA system (includes planting on raised beds) offers various opportunities for crop diversification. Cropping sequences/rotations and agroforestry systems when adopted in appropriate spatial and temporal patterns can further enhance natural ecological processes which contribute to system resilience and reduced vulnerability to yield reducing disease/pest problems. Limited studies indicate that a variety of crops like mustard, chickpea, pigeonpea, sugarcane, *etc.*, could be well adapted to the new systems with advantage.

Constratins in Adopting Conservation Agriculture Systems

Conservation agriculture poses a challenge both for the scientific community and the farmers to overcome the past mindset and explore the opportunities that CA offers for natural resources improvement. CA is now considered a route to sustainable agriculture. Spread of CA, therefore, will call for a greatly strengthened research and linked development efforts. Although significant successful efforts have been made in developing and promoting machinery for seeding of various crops in no till system, successful adoption of CA systems will call for greatly accelerated effort in developing, standardizing and promoting quality machinery aimed at a range of crop and cropping sequences, permanent bed and furrow planting systems, harvesting operations to manage crop residues, *etc.* Conservation agriculture systems imply that the whole range of practices including planting and harvesting, water and nutrient management, diseases and pest control *etc.* need to be evolved, evaluated and matched in the context of new systems.

Conclusions

CA involves the application of modern agriculture tools to improve production, enables not only the maximization of yields but also helps maintain the health and integrity of the ecosystem unlike the traditional systems which mainly aim to maximize yields often at the cost of the environment. Expansion of conservation agriculture can create a win-w*in situ*ation. Awareness creation, increased investment in research, policies and incentives could help promote CA and the spread of its benefits. CA must be brought to the main stream of crop management research and be closely linked with crop breeders and other agricultural disciplines for the development of cultivars, weed-pest disease management strategies.

REFERENCES

Kendall H W and Pimental D (1994). Constraints on the expansion of the global food supply. *Ambio* 23: 198-205.

Reicosky D C and Saxton K E (2007). Reduced environmrntal emissions and carbon sequestration pp: 52-58.

Vlek P L G (2008). The incipient threat of land degradation. *J of Ind. Soc of soil Science.* 56 (1): 83-87.

Conservation Agriculture Technologies *Pages* **15–36**
Editor: **Rahul Kumar, Abhinandan Singh and Pankaj Kumar Ojha**
Published by: **BIOTECH BOOKS, NEW DELHI**

2 Remote Sensing and GIS as a Tools for Natural Resource Management

Aditya Kumar[1] *and Abhinandan Singh*[2]

[1]***Department of Agricultural Chemistry and Soil Science,***
Udai Pratap Autonomous College, Varanasi – 221 002, Uttar Pradesh
[2]***Department of Agronomy, Dr Rajendra Prasad Central Agricultural University,***
PUSA, Samastipur – 848 125, Bihar

Remote sensing and Geographical Information System (GIS) provide abundant opportunity to monitor and manage natural resources at multi-temporal, multi-spectral and multi-spatial resolution. The advanced techniques like remote sensing, global positioning system and geographical information system can be of great use for natural resources assessment and management. These technologies have manifold applications in agriculture including crop discrimination, crop growth monitoring, stress detection, crop inventory, soil moisture estimation, computation of crop evapo-transpiration, site-specific management/precision agriculture, crop acreage estimation and yield prediction. Remote sensing and GIS can also be used very effectively in land use/land cover analysis as well as damage assessment because of drought, floods and other extreme weather events. In this chapter, we compile the various applications of remote sensing and GIS tools that can be used for natural resource management (agriculture, water, forest, soil, natural hazards).

Keywords: *Remote sensing, Geographical Information System (GIS), Natural resource management.*

INTRODUCTION

Natural resources are gift of nature used in the production of goods and services. It includes land, vegetation, air, water, wildlife and minerals. Apart from their fundamental importance in the economic development of many nations, natural resources are of ecological significance in protecting the environment and adding aesthetic values to the natural settings (Auty, 2007).Natural resource management involves manipulation of the resource to preserve or supply products on a sustained basis (Knight and Bates, 1995).Currently, there are many procedures used in the assessment of natural resources which include field-surveys, sampling, rating, ranking and assigning of scores to categorize the status of natural resources ecological unit based on environmental conditions and anticipated improvements (Carpenter *et al.*, 2009).By this way they are classified either as 'sustainable' or 'vulnerable' or 'endangered' resources. The aforementioned procedures lack objectivity necessitating compromise in thematic details. They are rudimentary cataloguing schemes that results in wrong classification of environmental performance indicators. Furthermore, they are expensive, time consuming and labor intensive. Apart of these limitations they pose difficulties in data integration and compatibility from different sources and techniques (Nykvist and Nilsson, 2009). It is, therefore, essential to adopt the technologies that will improve data quality and efficiently handle large amount of data by using modern techniques. Geospatial technology is a computer-based information system that acquires, integrates, stores, manages, analyses, interprets and displays output in two or three dimensions referenced to a particular location on the Earth's surface by geographic coordinates system (Elwood, 2008). Geo-informatics is a fast emerging science encompassing the modern tools of Remote Sensing, Geographic Information System (GIS), Global Positioning System (GPS) and simulation models. The combination of these technologies provide a cost effective means of acquiring high resolution real time data through remote sensing, data management and analysis through GIS and geo-referencing the ground truth data with GPS, putting all data in an information system and utilization of the information for a specific purpose. Utilizing geospatial technology for managing natural resources has a special importance when data is to be collected from some hilly and swampy terrains. They offer a rapid, affordable and powerful tool that has the ability to efficiently handle geospatial datasets, create maps, provide interactive user-friendly interfaces, transform spatial data according to users need, administer datasets from desktop through local network to global Internet facilities, visualize, analyse and interpret the interaction of spatially distributed natural resources in a timely, reliable, relational, and cost-effective way for valuable solutions (Descamps *et al.*, 2006).

According to Lillesand and Kiefer (1979) Remote sensing is defined as "the science and art of acquiring information (spectral, spatial, and temporal) about material objects, area, or phenomenon, without coming into physical contact with the objects, or area, or phenomenon under investigation". This technology measures Electro Magnetic Energy (EME) from the sun or emitted by the sensor itself (microwave sensor) which is reflected, scattered or emitted by different surface features on the earth. Remote Sensing employs passive and/or active sensors. Passive sensors are those, which sense natural radiations, either reflected or emitted from the earth. On the other hand, the sensors, which produce their own electromagnetic radiation, are called active sensors (*e.g.* LIDAR, RADAR). Detection and measurement of spectral signatures enable identification of the surface features by remote sensing satellites. It provides multi-spectral, multi-sensor and multi-temporal data which help in the generation of accurate, timely and cost effective information on natural resources. Remote sensing technology may be divided into three phases:

(i) Data collection from a sensor mounted on a platform *e.g.* a satellite

(ii) Data handling

(iii) Data interpretation which end up in producing some thematic maps of the investigated surfaces.

Data acquired by remote sensing is interpreted with the help of GIS software. A Geographical Information System is a system of hardware, software and procedures to facilitate the management, manipulation, analysis, modeling, representation, display of geo data to solve complex problems regarding planning and management of resources. Basically GIS is mapping software that links information about where things are with information about what things are like. Any GIS application contains five essential elements: data acquisition, preprocessing, data management, manipulation, analysis and product generation (Canter *et al.*, 1994). The geographical reference cartesian or latitude and longitude co-ordinate has proved to be an effective means of linkage of data sets and this principle, perhaps more than any other is the reason for the success of GIS. By performing query across the data sets of information at random more complex questions can be answered and a far broader range of problems can be solved. GIS are frequently used to answer many types of generic questions *e.g.* location, condition, pattern routing, modeling, *etc.* The major advantage of GIS is that it can read and analyze different layers of spatial information in the form of maps and satellite images easily and allows identifying the spatial relationships. There are many GIS softwares as per user's needs *e.g.* ERDAS, ArcView, ArcGIS, SWAT, which allow developers to add dynamic mapping, improves image quality, overlapping *etc.* (Ormsby *et al.*, 2010).

Remote Sensing and GIS used as efficient tool in many broad aspects of natural resource management like Agriculture, Soil Survey and Mapping, Land use and Land cover studies along with change detection studies, Environmental studies, Biodiversity management and Forest and Disaster management.

Application of Remote Sensing and GIS in Agriculture

Agriculture plays a dominant role in the economies of both developed and undeveloped countries. Satellite and airborne images are used as mapping tools to classify crops, examine their health, examine their viability, and monitor farming practices. Agricultural applications of remote sensing include crop type classification, crop condition assessment, crop yield estimation, mapping of soil characteristics, mapping of soil management practices, and compliance monitoring (farming practices). There has been increased emphasis on the potential utility of using remote sensing platforms to obtain real time assessments of the agricultural landscape. Application of Remote Sensing and GIS in Agriculture can be categorized under following headings:

A) Soil Management using Remote Sensing and GIS

Soil is a valuable non-renewable resource, which provides essential support to ecosystems and exists throughout the world in diverse behavior and properties. Remote Sensing GIS is used effectively in the following major domains of soil management.

1. Soil Moisture Stress Determination

Soil moisture stress management is one of the most important fields where we can choose remote sensing and GIS as important tools through the application of precision farming. The remote sensing techniques can be used in identifying and delineating areas of different grades of soil moisture whereas, the microwave remote sensing holds promise for sub soil moisture measurements. In the semi-arid and arid regions, judicious use of water can be possible through adaptation of precision technologies. For example drip irrigation coupled with information from remotely sensed data such canopyair temperature difference can be used to increase the water use efficiency by reducing the runoff and percolation losses (Das and Singh, 1989). Development in remote sensing data acquisition capabilities, data processing and interpretation of ground based, airborne and satellite observations have made it possible to couple remote sensing technologies and crop management systems to improve nutrient and water use efficiency. Mukherjee *et al.* (2014) standardized the spectral reflectance characteristics of wheat under water stress conditions. The spectral reflectance in the visible region was higher in water stressed crop than the non-stressed. The vegetation indices like NDVI, RVI, PVI and GI were found lower for stressed and higher for non-stressed crop.

2. Soil Degradation Estimation

Soil degradation (the decline in soil quality caused by human misuse) is one of the most important challenges facing mankind (Oldeman, 1988). It affects soil quality for agriculture and has implications for urban environment, pollution and flooding. The soil becomes less able to support plant and animal growth as there is a decline in levels of available moisture, and nutrients and biological activity. Acidification, salinity, organic carbon depletion, compaction, nutrient depletion, chemical contamination and erosion are all forms of soil degradation that can be brought about by inappropriate land use practices. Due to ever increasing pressure of population on land, the per capita arable land has been reducing. In the year 1986, the worldwide cropped area was 1.5 billion ha which was supporting the total world population of about 5 billion. The per capita arable land in 1986, thus works out to be 0.3 ha. With the increasing population pressure it has been progressively declining. By the year 2000, the per capita arable land area will decline to 0.23ha, and to 0.15ha by 2050 (Lal and Pierce, 1991). With only 55 per cent of the geographical area of the world, the developing countries carry 75 per cent of the world population which leads to over-exploitation of natural resources. Gopal Krishan *et al.* (2009) used vegetation cover, slope and erosion status derived from remote sensing data to delineate four major land degradation categories *viz.*, undegraded, moderately degraded, degraded and severely degraded. The potential utility of remotely reused data in the form of aerial photographs and satellite sensed data has been well recognized in mapping and assessing landscape attributes controlling soil erosion pattern (Pande *et al.*, 1992).

3. Soil Survey and Mapping

Soil survey provides an accurate and scientific inventory of different soils, their kind and nature, and extent of distribution so that one can make prediction about their characteristics and potentialities (Mandal and Sharma 2005). It also provides classification of soils and other properties in an area and geo-encoding such comprehensive information enhances the knowledge on natural resources for their proper planning and management. The traditional methods of gathering information are expensive and time consuming due to large number of observations. However, advances in computer and information technology have introduced new group of tools, methods, instruments and systems. Rapid developments in new technologies such as remote sensing (RS), GPS and GIS provide new approaches to meet the demand of resource planning and successfully used in studying the various aspects of soils in spatial and temporal domain (Shrestha 2006, Yeung and Lo 2002).The soil maps are required on different scales to meet the requirements of planning at various levels. The larger the scale used, more will be the information available and vice-versa. The coarse resolution data (spatial resolution 70m or more) from IRS LISS-I, AWiFs and LANDSAT-MSS sensors were useful to prepare soil maps on 1: 250,000 scale or smaller. To map soils on 1: 50,000 scale medium resolution data from LANDSATTM, IRS LISS-II and SPOT-MLA are employed. Satellite data from IRS-P6 (LISS-IV sensor), Cartosat-1 and Cartosat- 2 and IKONOS are now being employed for detailed characterization of soils on 1: 10,000 scale or larger (Dwivedi, 2001). The traditional method of soil analysis and interpretation are laborious, time consuming, thus becoming expensive hence, *kriging* and its variants have became widely recognized as an important spatial interpolation technique in land resource inventories (Hengl *et al.*, 2004). Remote sensing and GIS was successfully used for natural resource mapping and soil taxonomic study by Velmurugan and Carlos, (2009).

B) Water Management using Remote Sensing and GIS

Water as a resource is essential to support human existence. The availability of freshwater for human use has been declining over the years, whereas the demand of growing population is increasing. In this context, there is an urgent need to monitor and obtain a better understanding of its use, which will provide information that can assist towards the development of effective water management strategies and infrastructures. In the field of water resources (including drinking and irrigation water), there are several areas where geospatial technologies are being applied for scientific planning, management and monitoring. The two most important areas in this respect are: (i) Hydrological studies which include rainfall estimation, forecasting and monitoring, hydrological modeling, urban hydrology and water balance models, and (ii) Watershed conservation, planning and management, including watershed delineation, quantitative analysis of drainage network, watershed, geology, soil mapping, lake and reservoir sedimentation studies. About 210 thousands boreholes have been drilled annually with an average success rate of 85 per cent. A comparative study of failed hand pumps in several states was conducted by the UNICEF in 1995 to assess the impact of geophysical investigations. It was found that Tamil Nadu which used geophysical investigations had a success rate of 97 per cent, whereas states like Maharashtra and Madhya Pradesh, where adhoc drilling based on random selection or community convenience drilling was done, the rate of dry boreholes was as high as 40 per cent. Similarly, in Mizoram the introduction of geophysical surveying had made it possible to locate high yielding borewells where previously all efforts in this direction had been failed. The technology has picked up and currently almost all the states in India are using this valuable tool for groundwater investigations and also for locating rainwater harvesting structures. Remote sensing technologies would then be a promising tool to integrate with the hydrological models for getting continuous input data in data scarce regions. The launch of several Earth Observation (EO) sensors from advanced satellites provides world-wide continuous measurements on various hydrological components which are essential input data for hydrological modeling. Thus, satellite products and sophisticated computational techniques for the management of water can play an important role in following areas of water management.

1. Estimation of Evapo-transpiration (ET)

Estimation of evapo-transpiration (ET) is essential for water resource management such as water and energy balance computations, irrigation scheduling, reservoir water losses, runoff prediction, meteorology and climatology (Medina *et al.*, 1998). Estimation of spatial variability in evapo-transpiration is possible over a wide area by using remotely sensed information coupled with surface energy balance algorithms. The energy emitted from cropped area has been proven beneficial in assessing crop water stress as the temperature of most plant leaves are mediated by soil water availability and crop evapo-transpiration. Batra *et al.* (2006) estimated evaporative fraction (EF), defined as the ratio of ET and available radiant energy, by successfully using AVHRR and MODIS data. Several studies (Kite and Droogers, 2000; Loukas *et al.*, 2005; Eichinger *et al.*, 2006) have been conducted using more detailed hyper-spectral data, ancillary surface data and atmospheric data for improved spatial ET estimates. The availability of water, radiant energy and the removal of water vapour away from the surface are the major factors that control ET. However these factors in turn depend on other variables such as soil moisture, land surface temperature, air temperature, and vegetation cover, vapour pressure, and wind speed which may vary between regions, seasons, and time of day. Generally these factors are accounted for by using a combination of remote sensing data, ancillary surface data and atmospheric data for the estimation of ET values and has lead to extensive measurements of surface fluxes, meteorological and soil variables (Wang *et al.*, 2006). Batra *et al.* (2006) successfully estimated ET based on the extension of the Priestly-Taylor equation and the relationship between remotely sensed surface temperature and vegetation spectral indices. Folhes *et al.* (2009) employed Landsat imagery in conjunction with an evapo-transpiration model to measure water use levels in an irrigated area in the semi-arid northeast region of Brazil. Results showed that the combination approach of remote sensing and process modeling produced better predictability of water consumption in irrigated agriculture, and hence improved water resource management in irrigated areas. Ines *et al.* (2006) combined Landsat7 ETM+ images and derived distributed data such as sowing dates, irrigation practices, soil properties, depth to groundwater and water quality as inputs in exploring water management options in Kaithal, Haryana, India during 2000–2001 dry season. They revealed that under limited water conditions, regional wheat yield could improve further if water and crop management practices are considered simultaneously and not independently.

2. Watershed Management

Watershed management implies prudent use of all the natural resources to ensure optimum and sustained productivity. Development programs concerning optimum utilization of natural resources are being implemented on watershed bases. Watershed, as normally defined, is a natural hydrological entity that covers a specific area within which the entire rainfall runoff ultimately passes through a specific channel at a particular point. The information about the natural resources of the watershed is very important for watershed management planning. The various parameters of the watershed, *i.e.* stream network (drainage), physiography, land use, vegetation/forest cover and snow cover can be mapped and monitored using remote sensing data. This is essential for the conservation and management of water and land resources of watersheds for optimum productivity. In general, the remote sensing techniques in the field of watershed management is applied for:

a) Delineation and codification of watershed area
b) Watershed characterization and assessing watershed priority, evaluation of problems, potentials and management requirements of various watersheds
c) Erosion intensity mapping and identification of erosion prone areas
d) Soil, land use/land cover mapping
e) Drainage pattern mapping
f) Evolving water conservation strategies in a watershed
g) Selection of sites for the construction of check dams/reservoirs on streams or streamlets.
h) Suggesting sites for rain water harvesting structures
i) Evaluation and monitoring of the impact of the treatment

Thus, blending of remote sensing and GIS technologies has proved to be an efficient tool and have been successfully used by various investigators for water resources development and management projects as well as for watershed characterization and prioritization.

C) Crop Management using Remote Sensing and GIS

The role of Remote Sensing and GIS in crop studies can be broadly categorized into two groups - inventorying/mapping and management. While Remote Sensing data alone is mostly used for inventorying purposes (crop acreage estimation, crop condition assessment, crop yield forecasting, *etc.*) and management (cropping system analysis, precision farming *etc.*) In addition, it needs various other spatial agro-physical/environmental information has to be integrated with RS data, where the functionalities of GIS are used. Use of remote sensing (RS) data in making crop production forecast has been actively investigated in India and many other parts of the world. Monitoring Agriculture with Remote Sensing (MARS) programme, developed at the Joint Research Centre (JRC) for the General Directorate for the Agriculture of the Commission of European Communities (EEC), provides up-to-date agricultural information to European agricultural policy makers have demonstrated the integrated use of Remote Sensing, Geographical Information System (GIS) and historical database for improving cropland classification. Some of the specific activities in which Remote Sensing data and technology, alone or along with GIS are as follows:

1. Precision Agriculture

Precision agriculture is a production system that promotes variable management practices within a field, according to site conditions. These include the global positioning system (GPS), geographic information systems (GIS), yield monitoring devices, soil, plant and pest sensors, remote sensing, and variable rate technologies for applicators of inputs (Seelan *et al.*, 2003). Precision farming aims at reduced cost of cultivation, improved control and improved resource use efficiency through information received by the sensors fitted with the farm machineries. Variable rate technology (VRT) is the most advanced component of precision farming. Sensors are mounted on the moving farm machineries containing a computer which provides input recommendation maps and thereby controls the application of inputs based on the information received from GPS receiver (NRC, 1997). The area available for agriculture is getting reduced day-to-day because of rapid industrialization and urbanization. The problem in front of the modern day farmer is to produce more from less area. Hence, the grower has to treat different grids of its large field differently for input application. In recent years the expanded use of GPS and GIS has given rise to agricultural advances in spatial data management that have revolutionized the way many growers manage their fields (Anderson *et al.*, 1999). In near future, a farmer will go to the field with a spreader or combine equipped with a GPS that records positional information related to variable rate of fertilizer application or yield at harvest for his farm. This ability to monitor the variability in the field and then to pinpoint areas for input application using variable rate method is known as precision farming. Many growers in developed countries use this information for time and site specific field analysis with the assistance of farm management systems.

2. Crop Irrigation Demand Monitoring

Agriculture is the major consumer of water, utilizing more than 70 per cent of the global freshwater. Hence, the role of irrigation water plays a significant part in increasing land productivity. Land surface evapo-transpiration (ET) is one of the main components of the water balance that is responsible for water loss (Michailidis *et al.*, 2009) and it is of prime interest for environmental applications, such as optimizing irrigation water use, irrigation system performance, crop water deficit, *etc.* In this context, remote sensing technology has been emerged as an effective tool to monitor irrigated lands over a variety of climatic conditions and locations over the last few decades. It helps in determining when and how much to irrigate by monitoring plant water status, by measuring rates of evapo-transpiration and by estimating crop coefficients. The effective use of surface water and the monitoring of consumptive use of water using remote sensing techniques has been a topic of great interest for irrigation water policy makers.

3. Crop Production Forecasting and Acreage Estimation

Crop production forecasting comprises identification of crops, acreage estimation and forecasting their yield. Crop identification and discrimination are based upon the fact that each crop has a unique spectral signature. Typical spectral reflectance of a crop shows absorption due to pigments in the visible region, high reflectance in the near infrared region because of internal cellular structure of the leaves, and absorption in the short wave infrared region due to water content. Spectral index of the crop canopy (NIR/Red, Greenness, and NDVI) at any given point of time reveals the crop growth and its decay as affected by various factors in the time domain. Efforts have been made to develop a relationship between yield and spectral index using space borne data at maximum vegetative cover, and using different parameters of the growth profile like leaf area index (LAI). Spectral response of a crop canopy is influenced by (i) the leaf area index (LAI) and per cent ground cover, (ii) growth stage, (iii) differences in cultural practices, (iv) stress conditions and (v) canopy architecture. Background soil/water is an important influencing factor. Each crop has its own architecture, growing period *etc.*, thus enabling discrimination through RS data. If there are two crops with similar spectral signatures on a given date (confusion crops), multi-date data may be used to discriminate them. It has been observed that the ratio of near infrared to red radiance is a good indicator to red radiance is a good indicator of the vigour of the crop. Some of these properties are utilized in crop identification, yield forecasting and crop condition assessment.

4. Acreage Estimation

Identification and discrimination of various crops/land cover classes require quantitative use of subtle differences in their spectral data, and hence rely mostly on digital image processing techniques. The acreage estimation procedure broadly consists of identifying representative sites of various crops/land cover classes on the image based on the ground truth collected, generation of signatures for different training sites and classifying the image, using training statistics. Most of the work carried out so far has single-date data corresponding to the near maximum vegetative growth stage of the crop. District level acreage estimation has been generally carried out by analyzing the complete data. Estimation of crop acreage for large areas, like states require handling of a very large volume of data, larger efforts in ground truth data collection *etc.* In such case, sampling technique based procedures have been developed and successfully used. Based on the crop concentration statistics, agro-physical and/or agro-climatic conditions, the study area is divided into homogenous strata and sample segments from each stratum are analyzed. For digital data analysis, generally ten per cent of the total population is used. Appropriate statistical methods are employed to aggregate results at stratum and study area/state levels. In addition, national level forecast for wheat production in major wheat growing states of India has been made since 1995-96 season using multi-date IRS-1C/1D, WiFS and LISS-III data employing supervised hierarchical classification approach and agro-meteorological yield models. Availability of cloud free optical sensor data is a major problem especially during kharif season. The potential of microwave sensor data acquired in C-band at large incidence angles has been studied for acreage estimation and monitoring of rice crop during kharif (monsoon) season. Temporal data acquired during transplanting, vegetative and grain filling stages of rice had resulted in classifications of all types of lowland cultivars to the tune of 90 per cent accuracy. Many horticultural crops, *viz.* mango, coconut, areca nut, orange and banana could also be identified and their acreage estimated using high spatial resolution LISS-III data.

5. Crop Modeling

It is possible to combine crop models and remote sensing in the way of evaluate yield variables from remote sensed data for each time step in the model simulations, thus the use of remote sensing allows us to fill the missing model parameters during the recalibration in field scale (Batchelor *et al.*, 2002). Additionally, getting data from crop models in field scale -remote sense allows transferring the results from field scale to regional scale (Priya and Shibasaki, 2001). Many ways of using remote sensing data with crop models have been suggested (Wiegand *et al.*, 1986 and Dele´colle *et al.*, 1992). One way is estimating LAI (leaf area index) values by remote sensing for calibrating into the crop models. Other way is early estimates of the final yield but this method needs many remote sensing data during the growing season to use in crop models. Baret *et al.* (2006) combined remote sensing observation with crop models for providing stress quantification through assimilation approaches. Crop and soil model with GIS can be used to detect methane emission from fields (Matthews *et al.*, 2000) similarly it is used to estimate global food production and the impacts of global warming using GIS and crop model.

6. Crop Condition Assessment

Efforts have been made to develop procedure for crop stress detection and condition monitoring using satellite data at Space Applications Centre (SAC), Ahmedabad. Landsat MSS data has been used to detect the moisture stress in groundnut crop in Vanthali taluk of Junagadh district. Resource satellites like IRS or Landsat can provide detailed spatial information on the condition of the crop but the utility of such data to regional or large area crop condition study is restricted because of their comparatively long repetition cycles and the substantial volume of data. To handle the problem of crop condition assessment, Ajai and Sahai (1986) have proposed a hybrid approach, which uses both the comparatively high resolution earth resources satellite data together with the coarse-resolution meteorological satellite data (NOAA-AVHRR). Liu and Kogan (1996) have used the NDVI images generated from NOAA-AVHRR GVI (Global Vegetation Index) data to monitor large-scale drought patterns and their climatic impact on vegetation. Kogan (1994) has developed two indices, like. VCI (Vegetation Condition Index) and TCI (Temperature Condition Index) for monitoring drought. While VCI is the percentage of NDVI with respect to its maximum amplitude, TCI is the percentage in brightness temperature (derived from channel 4 of NOAA-AVHRR) with respect to its maximum amplitude. In India a National Agricultural Drought Assessment and Monitoring System (NADAMS) was initiated in 1986, using remote sensing data from the NOAA satellite and ground observations of rainfall and agricultural conditions.

7. Cropping System Analysis

In Mid 80's it was realized that component approach of crop research is not sufficient to sustain agriculture production level achieved during Green Revolution era. A system approach that takes into account the interaction of biotic and abiotic components of agriculture in time and space is considered as the viable alternative to sustainable agriculture production. This has given rise to cropping system approach which encompasses the study of cropping patterns and their management needs to derive maximum benefits from a given resource base under specific environmental conditions. A sustainable cropping system needs to be adopted keeping in view the available water, suitability to sustain a long term growth in productivity in a particular agro-climatic zone, keeping in view the domestic need, market demand and price. The scope of cropping systems study, include:

a) Describing the present cropping system
b) Monitoring the long-term changes in the cropping system
c) Evaluation of the long-term effects of present cropping system, and
d) Prescribing the alternate cropping system.

Cropping pattern of an area is the manifestation of the climate, the soil, available facilities like irrigation, fertilizer, mechanization *etc.* and the socio-economic factors. Changes in cropping pattern are also caused in response to change in one or more of the same factors. However, many times, the socio economic factors play a dominant role for adopting a new cropping system and thereby neglecting/degrading the parameters essential for agro-ecological balance. This information when integrated with other physical, environmental and economic resources, through GIS based simulation models, long-term performance of the cropping system can be evaluated. Investigations carried out at SAC, Ahmedabad using temporal remote sensing data from IRS sensors gave promising results for cropping system studies. Panigrahy and Sharma (1995) first generated crop rotation map using IRS data. Spatial database of parameters like crop area, cropping pattern, crop rotation, crop calendar, crop vigour *etc.* was created using multi temporal, mutispectral data. Panigrahy and Sharma (1997) while studying the crop rotation in Bardhaman district of West Bengal using multi-temporal IRS LISS-III data, concluded that the number of data sets required can be reduced by taking into account of crop calendars and including data of transitional phases. Using this methodology, Cropping System Analysis in Punjab was undertaken by Punjab Remote Sensing Centre, Ludhiana and Space Applications Centre, Ahmedabad under Jai Vigyan National Science and Technology Mission. Similar studies were undertaken for West Bengal and Assam states. Multi-date, multi-year and multi sensor satellite based remote sensing data along with various spatial and non-spatial collateral data was used to create spatial database and generate the cropping pattern and crop rotation maps for Punjab state. In addition, maps depicting cropping calendar and cropping intensity were also derived from remote sensing data. Various efficiency indices, such as Multiple Cropping Index (MCI), Area Diversity Index (ADI), Cultivated Land Utilization Index (CLUI), have been worked out for the year 1998-99 and 2004-05 using multi-date remote sensing data to characterize the cropping systems.

8. Monitoring of Vegetation Status

Remote sensing of soil and crop can be an attractive alternative to the traditional methods of field investigation because of the capability of covering large areas rapidly and repeatedly providing spatial and temporal information necessary for sustainable soil and crop management. The potential of remote sensing in agriculture is very high because of its ability to infer about soil and vegetation cover as a non-destructive mean. Numerous spectral vegetation indices (VIs) have been developed to characterize vegetation canopies. A most significant intellectual challenge to ecologists and bio-geographers is to understand spatiotemporal patterns of vegetation.

9. Physiological Studies

Remote sensing data have been exploited to estimate canopy characteristics by using empirical approaches based on spectral indices (Schlerf *et al.*, 2005). Different properties such as plant density, leaf area index, chlorophyll content *etc.* can be estimated using hyper-spectral imagery (Chun-Jiang *et al.*, 2006). These studies investigated the spectral reflectance properties of the plants, identifying key spectral wavebands related to specific plant physiological and structural characteristics, hence deriving sensitive vegetation spectral indices for their non-destructive estimation. Analysis of hyper-spectral remote sensing data has been carried out to estimate LAI for agricultural crops and forests. Schlerf *et al.* (2005) investigated several narrow and broad band vegetation indices in order to explore the possibility of hyper-spectral data to improve the estimation of biophysical variables such as LAI, canopy crown and crown volume when compared to multispectral analyses. The spectral and spatial information content of the satellite data was exploited to validate canopy reflectance models such as PROSPECT and SAILH (D'Urso *et al.*, 2004). Results obtained for the crops under investigation encourage the use of canopy reflectance models in the inverse mode in order to retrieve other vegetation parameters such as chlorophyll content, dry matter and canopy geometrical characteristics like mean leaf inclination angle. The accurate estimation of plant water status and plant water stress is essential for the integration of remote sensing into precision agricultural and forestry management. The potential to spectrally estimate plant physiological properties over relatively large areas, and to predict plant water status and plant water stress has been demonstrated for agricultural crops and forestry species.

10. Weed Identification and Management

Based on the variation in spectral reflectance characteristics of weeds and crops, remote sensing technology provides a means of identification of weed infestation in the crop stand and further aids in the development of weed maps by detecting the location of weeds within an agricultural field, so that site-specific/need based herbicide can be applied. Kaur *et al.* (2013) reported higher radiance ratio and NDVI values in solid stand or pure wheat and minimum under solid weed plots. It was observed that by using radiance ratio and NDVI, pure wheat can be distinguished from pure populations of *Rumex spinosus* beyond 30 DAS. Different levels of *Rumex* populations could be discriminated amongst themselves from 60 DAS onwards.

11. Pest Management

The remote sensing approach in assessing and monitoring insect defoliation has been used to relate differences in spectral responses to chlorosis, yellowing of leaves and foliage reduction over a given time period assuming that these differences can be correlated, classified and interpreted (Franklin, 2001). The range of remote sensing applications has included detecting and mapping defoliation, characterization of pattern disturbances *etc.* and providing data to pest management decision support system (Lee *et al.*, 2010). The possibility of forecasting and vulnerability of forest tress to insect defoliation has also been reported as tool for timely management. William *et al.* (1979) evaluated different types of vegetation indices on Landsat imagery acquired before and after defoliation to differentiate between healthy and unhealthy vegetation cover. Clerke and Dull (1990), determined the extent and severity of gypsy moth infestation in Virginia using imagery acquired by SPOT. De beurs and Townsend (2008) concluded that MODIS data represent an important tool for insect damaged defoliation and determination of vegetation indices in plot scale.

Application of Remote Sensing and GIS in Land Use and Land Cover Studies along with Change Detection Studies

Although the terms 'land cover' and 'land uses' are often used interchangeably, their actual meanings are quite distinct. Land cover refers to the surface cover on the ground, while land use refers to the purpose the land serves. The properties measured with remote sensing techniques relate to land cover from which land use can be inferred, particularly with ancillary data or a priori knowledge. Satellite remote sensing, in conjunction with geographic information systems (GIS), has been widely applied and been recognized as a powerful and effective tool in detecting land use and land cover change. It provides cost-effective multi-spectral and multi-temporal data, and turns them into information valuable for understanding and monitoring land development patterns. Satellite imagery has been used to monitor discrete land cover types by spectral classification or to estimate biophysical characteristics of land surfaces via linear relationships with spectral reflectances or indices (Steininger, 1996). GIS technology provides a flexible environment for storing, analyzing, and displaying digital data necessary for change detection and database development. Change detection by GIS is a process that measures how the attributes of particular area have changed between two or more time periods. Change detection often involves comparing aerial photographs or satellite imagery of the area taken at different times. It is useful in many applications such as land-use changes, habitat fragmentation, deforestation assessment, coastal change and urban sprawl. For example, Ochego (2003) used GIS to map forest loss and identify key factors involved in deforestation process. Laurance *et al.* (2001) also used GIS to predict regional-sclae changes in Brazillian Amazon. Prenzel and Treitz (2004) applied change detection techniques in watershed studies to enhance the capacity of local governments to sound environmental management. In mountainous region, difficult terrain and inaccessibility makes it almost impractical to obtain information required for efficient management of the natural resources with reasonable accuracy. Remote Sensing has proved its potential for providing such information. Ghosh *et al.* (1996) studied the land use changes in the Pranmati watershed in the catchment area of river Ganga in Chamoli district. Integration of remote sensing data with other spatial/non-spatial data was carried out using ARC/INFO software package. A simple classification technique was adopted for land cover/land use change analysis in relation to elevation, slope, aspect and bio-climatic classes. Suitability assessment of land where agricultural extension occurred between 1963 and 1993 was made using GIS software package.

Application of Remote Sensing and GIS in Environmental studies

Monitoring is the process of observing and determining qualitative and quantitative transformations that may be occurring and to predict their future trends. GIS presents the opportunity of undertaking effective graphical and numerical monitoring of environmental and natural resource issues. Globally, GIS has been heavily depended upon for environmental monitoring *e.g.* pollution monitoring (Briggs *et al.,* 2000) and monitoring land degradation and soil loss (Manoj, 2000). The ability of a GIS to seamlessly integrate data of varying structures (*i.e.* vector, raster, tabular) and of varying scales makes it key to the success of integrated natural resource monitoring efforts. Geospatial technologies used as efficient tools in following field of environmental studies.

1. Water Quality Assessment of Water Bodies

Remote sensing techniques can be used to monitor water quality parameters (*i.e.,* suspended sediments (turbidity), chlorophyll, and temperature). Optical and thermal sensors on boats, aircraft, and satellites provide both spatial and temporal information needed to monitor changes in water quality parameters for developing management practices to improve water quality. Remote sensing has been also used to measure chlorophyll concentrations spatially and temporally based on empirical relationships with radiance or reflectance (Ritchie *et al.,* 1994). The empirical relationships (algorithms) between the concentration of suspended sediments and radiance or reflectance for a specific date and site were developed to predict the water quality for several years (Ritchie and Cooper, 1991).

2. Environmental Pollution and Greenhouse Effects

The fast growth in population, urbanization, and change in land use pattern in developing countries has resulted in pollution in land, water, air and noise. GIS technique helps in monitoring Green House Gas emission from the construction activities. The maps generated from several sources could be overlaid to prepare the emission scenario and its impact on settlements (Denga *et al.,* 2008).

Application of Remote Sensing and GIS in Forest Management

Forest is a vital organ of our ecosystem. They impact human lives in several ways, despite of having huge importance the world forest has been declining at an alarming rate. Using Remote Sensing data and GIS techniques, a forest manager can generate information regarding forest cover; types of forest present within an area of interest, human encroachment extent into forest land/protected areas, encroachment of desert like conditions and so on. This information is crucial for the development of forest management plans and in the process of decision making to ensure that effective policies should put in place to control and govern the manner in which forest resources can be utilized. The suitability and status of sites/forest area for a particular species of wildlife can also be assessed using remote sensing data using multi criteria analysis. The digital analysis of the satellite images has the advantage of being fast and able to handle large amount of data whereas the visual image interpretation is based on the tonal variation of different features that is observed on satellite imagery.

Remote Sensing and GIS has proven to play a vital role in the following activities of forest management:

1. Resource Management

Collecting forest inventory data and monitoring changes are critical to forest management activities. GIS can be effectively used in the field of timber harvesting, silviculture and fire management activities, or predict fuel wood and other resource supplies. Other priorities, such as providing for wildlife habitat, ensuring recreation opportunities and minimizing visual impacts of harvesting, are also growing in importance. Some applications deal with single management issues, such as timber production, while others illustrate how a mix of management concerns can be integrated through the use of GIS, such as timber production combined with habitat protection.

2. Harvest Planning

Good forest management practice requires detailed planning of harvesting activities. Harvest planning activities include the identification of felling directions, extraction routes, depots and sensitive zones such as wetlands. Maps constitute a basic planning tool for these activities. Other tactical harvest planning functions utilize maps to identify planned felling over a number of years, and to consolidate felling areas and extraction routes thereby permitting the efficient use of harvesting equipment and other resources.

3. Fire Management

The effect of fire on forest resources is an important management concern. Management activities include fire prevention, wildlife control, prescribed burning, and post fire recovery actions. The modelling capabilities of GIS have been quite effective in this context. Forest fire managers have used GIS for fuel mapping, weather condition mapping, and fire danger rating. Forest fires have an important influence on the vegetation cover, animals, plants, soil, stream flow, air quality, microclimate, and even general climate. The loss of timber is obvious and so is the damage to life and property. The loss of recreation value of the forest and the destruction of wildlife habitat are also consequences of forest fires. The key to managing approved burning activities was the ability to anticipate fire behaviour after ignition.

4. Map Production

Forest managers require a wide variety of maps to assist with their daily activities. Plantation maps are most commonly used for location purposes and may contain additional useful information such as roads, rivers, compartment boundaries, planted species, and compartment size. Other features such as topographic features (contours), infrastructure, water points, fire breaks, neighbours and conservation areas may be also included in the map.

5. GIS for Strategic Planning and Modeling

Forest management planning involves making predictions about what the future forest will look like relative to alternative management activities. This ability is crucial to nearly all aspects of management forecasting, particularly long term wood and wildlife supply. GIS stores both the geographic and numerical structure of the forest stands and links that spatial database to the planning models. It allows the manager to effectively add both the important temporal and spatial dimensions to the management planning process. Within the limits of the inventory and model, the manager can then map what the forest will look like in 5,10, 25, or 100 years in the future.

6. Forest Change Detection

Forest change detection and monitoring has been mainly based on the comparing of two or more temporal sets of data or images, comparing old aerial photographs and present satellite imagery and comparing satellite imagery with existing old maps. The major limitation in using satellite imagery for change detection of forest cover is the spatial and temporal resolution of the satellite sensors.

Application of Remote Sensing and GIS in Biodiversity Management

Geospatial data can also be used in the management of flora and fauna within protected areas. Ground and aerial photographs, for instance, are essential in this practice. Aerial and satellite photographs can be used to determine the presence and distribution of vegetation within a protected area. These photos can also be used to determine the presence and distribution of sinvasive species within an ecosystem. This information is essential as it determines the amount of cover and food that is present, particularly for herbivores during various seasons of the year. Aerial photographs can be used to ease the process of counting during animal census activities. The stop capability of photographs eases this process. It is always essential for protected area managers to determine the population and distribution of various animal species within a protected area to ensure that they have enough food and water, to eliminate the chances of overstocking that might lead to soil erosion and to ensure that a balance within the ecosystem is arrived at. Geospatial data can also be used to show human encroachment into protected areas as well as animal activities outside protected areas. This data is critical especially in the process of resolving human/wildlife conflicts. Finally, the use of GPS technology can be applied to monitor the movement of endangered species as well as newly introduced species to determine their progress as well as protecting them.

Application of Remote Sensing and GIS in Disaster Management

Extensive multi-temporal spatial data is required for the management of natural disasters such flooding, earthquakes, volcanic eruptions and landslides. In this context satellite remote sensing is an ideal tool that offers information over large areas and at short time intervals, which can be utilized in various phases of disaster management, such as prevention, preparedness, relief, reconstruction, early warning and monitoring. Along with remote sensing, GIS techniques are required to handle huge spatial data sets and hence have been gaining importance in disaster management (Van Westen, 2000). GIS provides the data base from which the evidence left behind by disasters that have occurred in the past can be interpreted, and combined with the other information to arrive at hazard maps, indicating which areas are potentially dangerous. For example, landslide hazards have been estimated and mapped of using GIS-based statistical models. Floods have been predicted, warned against and mapped by combining morphological information derived from digital terrain models with hydrological models. Lava flow pathways have been simulated and volcanic hazard maps have been developed using GIS techniques (Carrara *et al.*, 1999). In most of these studies, GIS and Remote sensing have been used to develop suitable disaster management strategies and occupational frameworks for their monitoring, assessment and mitigation.

Conclusion

Remote sensing and Geographic Information Systems (GIS) have greatly expanded opportunities for data integration, data analysis, modeling, and map production.With the rising pressure on natural resources due to the increasing human population, remote sensing and GIS can be used to manage these precious limited resources in an effective and efficient manner. Multi-temporal satellite data help to delineate the various change of the earth surface. Remote sensing has progressively expended applications in various fields of natural resource management such as Agriculture, Environmental studies, Forest and Bio-diversity management and Disaster management. Because of its varied applications and ability to allow users to collect, interpret, and manipulate data, Remote Sensing and GIS has become a useful tool for natural resource managers.

REFERENCES

Ajai and Sahai, B. 1986. Drought detection and quantification by remote sensing. *Jal Vigyan Sameksha* **1**: 138-152.

Anderson, J.E. 1999. Remote Sensing and precision agriculture: Ready for harvest or still maturing? *Photogrammetric Engineering and Remote Sensing* **65** (10): 1118-1123.

Auty, R.M. 2007. Natural resources, capital accumulation and the resource curse. *Ecological Economics* **61** (4), 627–634.

Baret, F., Houles, V. and Guérif, M. 2007. Quantification of plant stress using remote sensing observations and crop models: the case of nitrogen management. *Journal of Experimental Botany* **58** (4), 869-880.

Batchelor, W.D., Basso, B., and Paz, J.O. 2002. Examples of strategies to analyze spatial and temporal yield variability using crop models. *Eur. J. Agron.* **18**, 141–158.

Batra, N., Islam, S., Venturini, V., Bisht, G. and Jiang, L. 2006. Estimation and comparison of evapotranspiration from MODIS and AVHRR sensors for clear sky days over the Southern Great Plains. *Remote Sens Environ* **103:** 1-15.

Briggs, D. J., de Hoogh, C., Gulliver, J., Wills, J., Elliott, P., Kngham, S., and Smallbone, K. 2000. A regression-base method for mapping traffic-related air pollution: application and testing in four construting urban environments. *The Science of Total Environment* 253, 151-167.

Canter, L.W., Choudhury, A. K. and Vieux, B. E. 1994. Geographic Information Systems: A Tool for Strategic Groundwater Quality Management, *Journal of Environmental Planning and Management.* Vol 37: 3, pp. 251-266.

Carpenter, S.R., Mooney, H.A., Agard, J., Capistrano, D., DeFries, R.S. and Dý´az, S. 2009. Science for managing ecosystem services: beyond the Millennium Ecosystem Assessment. *Proceedings of the National Academy of Sciences of the United States of America* 106 (5), 1305.

Carrara, A., Guzzetti, F., Cardinali, M. and Reichenbach, P. 1999. Use of GIS technology in prediction of monitoring of landslide hazard. *Natural Hazards* 20,117-135.

Chun-jiang, Z., Ji-hua, W., Liang-yun, L., Wen-Jiang, H. and Qi-Fa, Z. 2006. Relationship of 2100-2300 nm spectral characteristics of wheat canopy to leaf area index and leaf N as affected by leaf water content. *Pedosphere* **16:** 333-38.

Clerke, W. and Dull, C. 1990. Evaluating the utility of SPOT digital imagery for delineating and categorizing gyspy moth defoliation. In: *Protecting Natural Resources with Remote Sensing,* The Third Forest Service Remote Sensing Application Conference, 9-13 April, American Society of Photogrammetry and Remote Sensing, Tucson, Arizona, pp. 22-32.

D'urso, G. D., DiniI, L., Vuolo, F., Alonso, L. and Guanter, L. 2004. Retrieval of leaf area index by inverting hyperspectral multiangular CHRIS PROBA data from SPARC 2003. *Proc. 2nd CHRIS Proba Workshop.* 28 to 30 April, ESA/ESRIN, Frascati, Italy.

Das, D.K. and Singh,G. 1989. Estimation of evapotranspiration and scheduling irrigation using remote sensing techniques. Proc. *Summer Inst. On agricultural remote sensing in monitoring crop growth and productivity,* IARI, New Delhi, 113-17.

De beurs, K. and Townsend, P. 2008. Estimating the effect of gypsy moth defoliation using MODIS. *Remote Sens Environ.***112**: 3983-90.

Dele´colle, R., Maas, S.J., Gue´rif, M. and Baret, F., 1992. Remote sensing and crop production models: present trends. 1991/01/14-18. ISPRS *J. Photogram. Remote Sensing* (NLD) **47** (2–3), 145–161.

Deng, J.S., Wang, K., Deng, Y. H. and Qis, G. J. 2008. PCA-based land use change detection and analysis using Multi-temporal and multisensory satellite data', *International Journal of Remote Sensing* **29**(16), 4823-4838.

Descamps, G., Therrien, P. and Therrien, R. 2006. An interactive and open approach for the analysis and diffusion of geoscientific data. *Computers and Geosciences* **32** (5), 643–655.

Dwivedi, R. S. 2001. Soil resource mapping: a remote sensing perspective. *Remote Sensing Reviews* **20**: 89–122.

Eichinger, W. E., Cooper, D. I., Hipps, L. E., Kustas, W. P., Neale, C.M.U. and Prueger, J.H. 2006. Spatial and temporal variation in evapotranspiration using Raman Lidar. *Adv Water Resour* **29:** 369-81.

Elwood, S. 2008. Geographic Information Science: new geovisualization technologies–emerging questions and linkages with GIScience research. *Progress in Human Geography* (33), 256–263.

Folhes, M.T., Renno, C. D. and Soares, J.V. 2009. Remote sensing for irrigation water management in the semiarid Northeast of Brazil. *Agric Water Manage* **96**: 1398-1408.

Franklin, S. 2001. *Remote Sensing for Sustainable Forest Management*. Lewis publisher, Boca Raton, Florida 407 p.

Ghosh, S., Sen, K.K., Rana, U, Rao, K.S. and Saxena K.G. 1996. Application of GIS for landuse/land covers change analysis in a mountainous terrain. *J. Indian Soc. Remote Sensing* **24** (3): 193-202.

Gopal Krishan., Kushwaha., S.P.S. and Velmurugan, A. 2009. Land Degradation Mapping in the Upper Catchment of River Tons *J. Indian Soc. Remote Sens* **37**: 49–59.

Hengl, T., Heuvelink, G.B.M. and Stein, A. 2004. A generic framework for spatial prediction of soil variables based on regression-kriging. *Geoderma* **120** (1-2): 75-93.

Ines, A.V.M., Honda, K., Das, Gupta A., Droogers, P. and Clemente, R. S. 2006. Combining remote sensing-simulation modeling and genetic algorithm optimization to explore water management options in irrigated agriculture. *Agric Water Manage* **83**: 221- 32.

Kaur, R., Jaidka, M. and Kingra, P. K. 2013. Study of optimum time span for distinguishing *Rumex spinosus* in wheat crop through spectral reflectance characteristics. *Proc. Natl. Acad. Sci., India, Sect. B Biol. Sci.* (Published online: 29 October, 2013).

Kite, G. and Droogers, P. 2000. Comparing evapotranspiration estimates from satellites, hydrological models and field data. *J Hydrol* **229:** 3-18.

Knight, R.L., Bates, S.F. 1995. A New Century For Natural Resource Management, Le Hegarat Mascle, S R Seltz, L Hubert Moy S Corgne, N Stach Performance of change detection using Remotely sensed data and evidential fusion: Comparison of three cases of application. *Int. J. Rem. Sensing* 27, 3515 – 3522 (2006).

Kogan, F.N. 1994. Application of vegetation index and brightness temperature for drought detection. *Advances in Space Research* **15** (11): 91-100.

Lal, R. and Pierce.1991.The vanishing resource. In Soil Management for Sustainability. Soil and Water Conservation Society, USA. pp. 1-5.

Laurance, W.F., Albernaz, A.K.M., Schroth, G., Fearnside, P., Delamonica, P. and Barber, C.2001. The future of the Brazillian Amazon Sceience, 291, 438-439.

Lee, W., Alchanatis, V., Yang, C., Hirafuji, M., Moshou, D. and Li, C. 2010. Sesning technologies for precision specialty crop production. *Computer and Electronic in Agriculture* **74**: 2-33.

Lillesand, T. M. and Kiefer, R. W.1979. Remote Sensing and Image Interpretation, John Wiley and Sons.

Liu, W.T. and Kogan, F.N. 1996. Monitoring regional drought using the vegetation condition index. *International Journal of Remote Sensing* **17** (14): 2761-2782.

Loukas, A., Vasiliades, L, Domenikiotis, C. and Dalezios, N.R. 2005. Basin wide actual evapotranspiration using NOAA/AVHRR satellite data. *Phys Chem Earth* **30:** 69-79.

Mandal, A. K. and Sharma, R. C. 2005. Computerization database on salt affected soils of Haryana state. *Journal of the Indian Society of Remote Sensing* **33**(3): 447-455.

Manoj, K. J. 2000. Estimation of soil erosion and sediment yield using GIS. *Journal of Hydrological Sciences* **45**(5), 771-786.

Matthews, R. B., Waamann, R. and Arah, J. 2000. Using a crop/soil simulation model and GIS techniques to assess methane emissions from rice fields in *Asia. I. model* development. *Nutrient cycling in agroecosystems* **58**: 141-159.

Medina, J. L., Camacho, E., Reca, J., Lopez, R. and Roldan, J. 1998. Determination and analysis of regional evapotranspiration in Spain based on remote sensing and GIS. *Phys Chem Earth* **23:** 427-32.

Michailidis, A., Mattas, K., Tzouramani, I. and Karamouzis, D. 2009. A Socioeconomic Valuation of an Irrigation System Project Based on Real Option Analysis Approach, *Water Resources Management* **23** (10).

Mukherjee, A., Singh, S. P. and Kingra, P. K. 2014. Spectral reflectance characteristics of wheat (*Triticum aestivum* L.) under stressed and non-stressed conditions. In Proceedings of Internationl Symposium on *"New Dimensions in Agrometeorology for Sustainable Agriculture* (*NASA-2014*) held at GBPUAT, Pantnagar during 16-18 October, 2014.

NRC 1997. Precision Agriculture in the 21st Century Geospatial and informationtechniques in crop management. National Academy Press, Washington DC. 149p.

Nykvist, B. and Nilsson, M. 2009. Are impact assessment procedures actually promoting sustainable development? Institutional perspectives on barriers and opportunities found in the Swedish committee system. *Environmental Impact Assessment Review* **29** (1), 15–24.

Ochego, H. 2003. Application of remote sensing in deforestation monitoring: a case study of the Aberdares (Kenya). 2nd FIG Regional Conference, Marrakech, Morocco.

Oldeman, L R., Hakkeling, R.T.A., and Sombroek, W.G. 1991. World map of the status of human-induced soil degradation: an explanatory note. (2nd edition) Wageningen: ISRIC.

Ormsby, T., Napoleon, E. J., Burke, R., Groessl, C. and Bowden, L. 2010.Getting to Know ArcGIS Desktop, 2nd edition. ESRI Press.

Pande, L.M., Prasad, J. Saha, S.K. and Subrahmanyam C. 1992. Revies of remote sensing application to soils and agriculture.

Panigrahy, S. and Sharma, S.A. 1995. Approach to map crop rotations using temporal IRS LISS-I data: A case study for Bardhaman district (West Bengal). Proceedings of National Symposium on Remote Sensing of Environment with Special Emphasis on Green Revolution, Ludhiana. Nov. 22-24, 1995 pp. 35-39.

Panigrahy, S. and Sharma, S.A. 1997. Crop rotation mapping using multidate IRS digital data. *International Society of Photogrammetry and Remote Sensing Journal of Photogrammetry*. **52**: 85-91.

Prenzel, B., Treitz, P. 2004. Remote sensing change detection for a watershed in north Sulawesi, Indonesia, *Progree in Planning* 61, 349-363.

Priya, S., and Shibasaki, R. 2001. National spatial crop yield simulation using GIS-based crop production model. *Ecological Modelling* **136** (2), 113-129.

Ritchie, J.C., and Cooper, C.M. 1991. An algorithm for using Landsat MSS for estimating surface suspended sediments, *Water Resources Bulletin* **27**: 373–379.

Ritchie, J.C., Schiebe, F.R., Cooper, C.M. and Harrington, J.A. Jr. 1994. Chlorophyll measurements in the presence of suspended sediment using broad band spectral sensors aboard satellites, *Journal of Freshwater Ecology* **9** (2): 197–206.

Schlerf, M., Atzberger, C. and Hill, J. 2005. Remote sensing of forest biophysical variables using HyMap imaging spectrometer data. *Remote Sens Environ* **95:** 177-194.

Seelan, S. K., Laguette, S., Casady, G. M., and Seielstad, G. A. 2003. *Remote sensing* applications for precision agriculture: A learning community approach. *Remote Sensing of Environment* **88** (1), 157-169.

Shrestha, H. L. 2006. Using GPS and GIS in participatory mapping of community forest in Nepal. *Electronic Journal of Information Systems in Developing Countries* **25**(5): 1-11.

Steininger, M.K. 1996. Tropical secondary forest regrowth in the Amazon: age, area and change estimation with Thematic Mapper data. *International Journal of Remote Sensing* **17**, 9–27.

Van Westen, C.J. 2000. International Archives of Photogrammetry and Remote Sensing. Vol. XXXIII, Part B7. Amsterdam.

Velmurugan, A. and Carlos, G. G. 2009. Soil Resource Assessment and Mapping using Remote Sensing and GIS. *J. Indian Soc. Remote Sens.* **37**: 537–547.

Wang K, Li, Z. and Cribb, M. 2006. Estimation of evaporative fraction from a combination of day and night land surface temperatures and NDVI: A new method to determine the Priestly-Taylor parameter. *Remote Sens Environ* **102:** 293-305.

Wiegand, C. L., Richardson, A. J., Jackson, R. D., Pinter, P.J., Jr., Aase, J.K., Smika, D.L., Lautenschlager, L.F. and McMurtrey, J.E., III. 1986. Development of agrometeorological crop model inputs from remotely sensed information. IEEE Trans. *Geosci. Remote Sens* GE-24: 90-98.

Williams, D., Stauffer, M. and Leung, K. 1979. A forester's look at the application of image manipulation techniques to Landsat data. In: Symposium on Remote Sensing for Vegtation Damage Assessment, February 14-16, Washington, The Society, Falls Church, VA, pp. 221-29.

Yeung, A. K. W. and Lo, C. P. 2002. Concepts and Techniques of Geographic Information Systems. *International Journal of Geographical Information Science* **17**(8): 819-820.

Conservation Agriculture Technologies *Pages* **37–46**
Editor: **Rahul Kumar, Abhinandan Singh and Pankaj Kumar Ojha**
Published by: **BIOTECH BOOKS, NEW DELHI**

3 Role of Extension in Resource Conservation Technology in Agriculture

Abul K. Azad

Ph.D. Scholar, Dairy Extension Division
ICAR-National Dairy Research Institute, Karnal – 132 001, Haryana

Resource conservation for the future generation is the prime priority for human being. If we utilize all the material whatever existing today with us without concerning the future aspect then it may be the most foolish things happened to be in the earth. Resource conservation is necessary not only for what we will be eating, cooking, utilising in near future but it has direct relation with the existence of our planet called Earth. The world population by 2050 will be 9 billion and India alone will have 1.4-1.7 billion. There will be huge hungry people in near future,food and feed production there must be increase 60 per cent to mitigate the demand (FAO, 2013). Therefore the resource conservation in agriculture is the call of nature to safeguard the flora, fauna and the Earth. However, the priority goal of resource conservation technology in agriculture is the long term sustainable crop production which mitigate the present generation needs as well as near future demand. The subject domain extension plays a vital role in this context of resource conservation agriculture. Extension act as both push factor and pull factor for this aspect.The domain extension gives training, create awareness, dissemination of low input technology, develop expertise, and motivate stakeholders to go for conservation agriculture. The remarkable organization/institute which are comes under the realm of extension to support participatory technology generation, dissemination of cost effective technology, decision making, weather information, marketing information and agricultural input supply with main vision to conserve and protect the natural resources without compromising future demand are Agricultural Technology Management Agency, Agricultural Technology Information Centre, Krishi Vigyan Kendra and various Information and Communication Technology tools.

Resource Conservation Technology in Agriculture

Before going to the technological parts of the chapter, we must know Resource Conservation Technology (RCT) in agriculture? Resources are anything which are useful for man-kind as well for animals. Resources broadly classified as man-made and natural. The man- made resources includes Fan, Chair, Electricity *etc.* whereas natural resources are Air, Water, Sunlight, Natural gas *etc.* Conservation of resources is the ethical use and protection of valuable resources, such as trees, minerals, wildlife, water and others. It focuses on maintaining the natural world in order to protect the sources of resources. Technology is defined as the application of scientific knowledge for practical purposes. Therefore the overall definition of RCT in Agriculture is refer as "the conservation of resources by application of scientific knowledge and management in agriculture to mitigate the future food and nutritional crises."

On account of above context, what is/are the role playing by theExtension functionaries to conservethe resources are briefly discussed.

Genesis of Extension and Historical Background

The extension genesis have long back roots. Extension is simple whatever technology we have let's pass it on to needy people. The people word includes farmers, entrepreneurs, women, youth, dairy stakeholder *etc.*According to Ensminger (1957), extension is education and that its purpose is to change attitudes and practices of the people with whom the work is done. Another way we can say that it is the system of working with farmers.

Knowledge Resource Person	→	End users	=	Extension

The word extension is derived from Latin word ***Extensio***which means stretching out. The term first originated in 1866 at Englandandthe word "extension" was coined by **McCarthy**. In 1873 "Extension education" term first used by Cambridge University, in 1876 by London University and in 1878 by Oxford University but this term gave more practical shape at field level by **J.P. Leagans** that is why he is known as ***Father of Extension.*** In Indian Context, **K. N Singh** is called as the father of Extension.

Some of the landmark historical acts such as Land Grant- The Morrill Act of 1862 about establishment of at least one college in each state with theObjective to teach agriculture and the mechanical arts without excluding other scientific and classical studies. The Hatch Act of 1887 recommended to create an experiment/ Research station at every land-grant college.In 1890 - second Morrill Act the funds were provided to sustain these colleges. Under the Act of Smith-Lever Act-1914 the Cooperative Extension Service has been set up. Another notable act *i.e.* Smith-Hughes Act 1917 was promoted vocational agricultural training to the farmers.

The historical background of extension in Indian context are the formation of Department of Revenue, Agriculture and commerce started functioning during

1871. Due to the series of famines occurred from 1875-1901; The British government was compelled to set up some commission. The agriculture became an independent branch in 1881.Imperial Agricultural Services was constituted in 1906 which was later ceased in 1926. Govt. of India, 1919 made the Agricultural Development become a state subject to uplift the poor farmers. In 1928 on the recommendation of Royal commission on Agriculture, the formation of ICAR took place in 16 July, 1929. In 1952, Grow More Food Enquiry committee recommended to set up an extension organisation in rural areas later it took the shape of Community Development Programme (1952) and National Extension Service (1953). The establishment of first State Agricultural University in 1960 at Pantnagar on land grant pattern.Last but not the least the creation of Department of Agriculture and Research Education was set up in 1973 at New Delhi.

How Extension is Helping To Conserve Resources in Agriculture?

The subject domain "Extension" helps to conserve the natural resources mainly in three way:

1. Research *e.g.* research on Climate-Smart Agriculture, Climate-Smart Livestock *etc.*
2. Extension *e.g.* Providing Advisory Services, Input supply,ToT, ICTs *etc.*
3. Teaching/Education *e.g.* Training on need basis, Farmer Field School, Demonstration *etc.*

Four Transfer of technology project of ICAR which comes under the domain of Extension.

National Demonstration

This programme was launched in the year of 1964 on major food crops production.The main rationale of this programme is that the agricultural scientist demonstrate first time before handover the technology to the farmers. Initially AICRP on Maize 1957 was done. In this programme there was specific yield target and there was no separate control plot. The operational area for demonstration was a one hectare. It was conducted with the local extension workers.

The specific objectives of national demonstrations, according to Prasad, Choudhary and Nayar (1987), were as follows:

1. To demonstrate convincingly to the farmers and extension personnel for adopting the technologies by considering locally available of resources.
2. To fully exploit these demonstrations for the purpose of training farmers and field extension functionaries in improved cultivation practices.
3. To provide the research workers an opportunity to get first-hand knowledge of the problems being faced by the farmers.
4. To determine the income and employment generation potentialities of the crops/subjects under demonstration, and educate the farmers and extension agents about them.

Operational Research Project

The ORPs were initiated in 1974-75. The main aimed at disseminating the proven technology in a discipline/area among farmers on a watershed basis, covering the whole village or a cluster of villages, and concurrently studying constraints (technological, extension or administrative) as barriers to the rapid spread of improved technical know-how [Prasad, Choudhary and Nayar, 1987]. This particular programme identified two types of problem facing by the farmers; one is agricultural related and another one is total resources development especially in the watershed areas.

The specific objectives

1. To assessment, accept and demonstrate the new agricultural technology on farmers' fields
2. To define the technology in terms of economic point of view of the farmers
3. To reveal the constraints of technology and farmers.
4. To demonstrate group action as a method of popularizing the modern technologies at a faster rate.

Krishi Vigyan Kendra (Farm Science Centre)

KVK is a Kendra (Centre) which is being provide the need based and skill oriented vocational training to the farmers as well as field level extension workers. It was first established in 1974 at Pondicherry under Tamil Nadu Agricultural University. Initially it was set up on the basis of backwardness of the geographical area such as hilly areas, drought prone areas, forest areas, coastal areas, flood prone areas, and areas dominated with tribal farmers, weaker sections, small farmers and landless labourers.

The three fundamental principles, *viz.*:

1. Agricultural production as the prime goal,
2. Work-experience as the main method of imparting training, and
3. Priority to weaker sections of the society, are the backbone of the KVK programme.

The main objective of the KVK is to provide a strong training support for bringing about production breakthrough in agriculture.

The mandate of KVK are Three folded [Mathur, 2011]

1. Technology assessment and Refinement
2. Training
3. Knowledge and Resources Centre of Agricultural Technology

But initially when it was established the mandates were **Vocational Training** in 1974 and **On-Farm Testing and Demonstration of Technology** during 1990s.

The specific objectives are as follows:

1. Strategy and comportmentassessment of the functioning area to formulate the resource inventory with special reference to identify the training needs of the farming community.
2. Strategy and comportment production oriented, need-based, short and long duration training courses both on the campus and off campus with priority on the weaker and poorer sections.
3. Organize Farm Science Clubs, SHGs, Women Empowerment Cell in villages to inculcate in the younger generation a liking for and an interest in agricultural and allied sciences
4. Develop and maintain the campus farms and demonstration units on scientific lines as the facilities for providing work- experience to the trainees as also dissemination of the latest technical knowhow.

All ICAR-TOT Programmes merged with KVK in 1st April, 1992.

Lab to Land Programme

The Lab to Land Programme (LLP) was launched by the ICAR in 1979 as a part of its Golden Jubilee celebrationwith 75,000 farm families over the whole country. The overall objective of the programme was to improve the economic condition of the small and marginal farmers and landless agricultural labourers, particularly scheduled castes and scheduled tribes, by transfer of improved technology developed by the agricultural universities, research institutes *etc.*

The specific objectives of the lab to land programme, according to Prasad, Choudhary and Nayar (1987) were:

1. Identification of resources of the selected farmers and landless agricultural labourers
2. Development of low input and maximum output technology for the farmers
3. Create linkage between the different stakeholders, Scientists and Institutions like NGOs
4. Organizing need based training for the farmers
5. Make awareness among the farmers about latest technology which is/are cost-effective and easy to adopt.

Another decentralised extension system to safeguard the farmers on conservation of resources in agriculture is **Agricultural Technology Management Agency (ATMA)**. It is a district level autonomous institution and bottom-up approach. It was established under the Innovation Technology Dissemination of National Agricultural Technology Project in 1999. It is jointly supported by National Institute of Agricultural Extension Management (MANAGE), State Agricultural Management and Training Institute (SAMETI) and Krishi Vigyan Kendra (KVK). The project activities are monitored by the Governing Body and District Magistrate/ Collector is the Chairman of the ATMA Governing Body.

There are two bodies created at the Block Level under Farm Information and Advisory CentreFIAC *i.e.* Farmer Advisory Committee which look after the Block Action Plan and Block Technology Team which is used to prepare the details of extension activities to be undertaken.

Farmers Interest Group (FIG) or Farmers Organization mainly created at village level to promote technology generation with their locally available resources and dissemination to the market driven farmers.

Agricultural Technology Information Centre

There are various technologies developed and developing by the various ICAR Institutes, State Agricultural Universities and Central Agricultural Universities for the betterment of the farmer but there was no common centre for all the technologies developed were stored together and provide it to the end users mostly on paid basis as well as free of cost in a "single window" delivery system. By keeping in view point, the establishment of Agricultural Technology Information Centre (ATIC) come up on. The functions of ATIC at institute level can be divided into Products, Services and Information.

Products	*Services*	*Informational*
Value Added Products made by the institutions Genetically improved Planting materials Agro-equipment's developed by the institution or any other	Soil Testing Disease identification and management Consultancy on animal diseases Testing irrigation water toxicity Breeding, Feeding and Management of animal	Leaflet, Pamphlet, Newsletters, Farm Magazine Audio-Visual Aids, Folder *etc.*

The Farmer FIRST Programme (FFP)

The Farmer FIRST is a new initiative launch during 12th FYP. It is a concept of ICAR where farmers stand in a centric role for research problem identification, prioritization and conduct of experiments and its management in farmers' conditions. The focus is on farmer's Farm, Innovations, Resources, Science and Technology (FIRST). In Indian context, 'enriching knowledge' and 'integrating technology' fulfil the meaning of Farmer FIRST. Enriching knowledge signifies the need for the research system as well as farmers to learn from each other in context to existing farm environment, perception of each other and interactions with the sub-systems established around. Technology integration is looked from the perspective that the scientific outputs coming out from the research institutions, many times do not fit as such in the farmers' conditions and thus, certain alterations and adaptations are required at field level for their acceptance, adoption and success.

Information and Communication Technology

Today's world is the era of information, Information and Communication Technology became the part and parcel of every aspect. The ICT tools can be applied for providing accurate, timely, relevant information and services to the farmers, thereby facilitating an environment for more remunerative agricultureand conservation of natural resources. Asper extension is concern the Information Technology can assistance here, by empowering extension workers to gather, store, retrieve and disseminate a broad range of information needed by farmers, thus transforming them from extension workers into knowledge workers (Saravana, 2010). Some of the ICT tools which are changing the agricultural scenario as well as helping in conservation of natural resources in agriculture are:

A) Kissan Call Centre

The "Kissan Call Centre" is one of the information dissemination system under the domain of extension launched by the Ministry of Agriculture and Farmers Welfare on January 21, 2004. The primary objective of this KCC is to answer farmers' queries on a telephone call in their own dialect. The queries answers are given in 22 local languages. The nationwide toll free number 1800-180-1551 has been allotted to the Kissan Call Centre. The number is accessible through mobile phones and landlines of all telecom networks including private service providers. These call Centres are working in 14 different locations covering all the States and UTs.

Call Center services are available from 6.00 am to 10.00 pm on all seven days of the week at each KCC location. Kisan Call Centre agents known as Farm Tele Advisor(FTAs), are graduates or above (*i.e.* PG or Doctorate) in Agriculture or allied (Horticulture, Animal Husbandry, Fisheries, Poultry, Bee-keeping, Sericulture, Aquaculture, Agricultural Engineering, Agricultural Marketing, Bio-technology, Home Science *etc.* and possess excellent communication skills in respective local language. Queries which cannot be answered by Farm Tele Advisor (FTAs) are transferred to higher level experts in a call conferencing mode. These experts are subject matter specialists of State Agriculture Departments, ICAR and State Agricultural Universities [ICAR 2016].

B) mKisan

This is the SMS portal for farmers which is used to give information, services and advisories to the farmers by SMS in their own language, preference of agricultural practices and location. This Portal was inaugurated by the Hon'ble President of India on July 16, 2013 and since its inception nearly 50 crore messages or more than 152 crore SMSs have been sent to farmers throughout the length and breadth of the country [Mondal 2014].

C) KRISHI

Knowledge based Resources Information Systems Hub for Innovations (KRISHI) in agriculture, is an initiative of Indian Council of Agricultural Research (ICAR) to bring its knowledge resources to all stakeholders at one place. The portal is being developed as a centralized data repository system of ICAR consisting of Technology, Data generated through Experiments/Surveys/Observational studies, Geo-spatial data, Publications, Learning Resources *etc.*

A large number of web-based transactional services are being provided to farmers in various States and also by organisations/departments of the Government of India. These range from issuance of soil health cards, provision of subsidized agricultural inputs and analysis of pests/diseases *etc.* Quite a few such services are in pipe-line under National e-Governance Plan-Agriculture or otherwise.

D) Agmarknet

This initiative provides daily market price and arrival information in respect of 300 commodities and 2000 varieties in eight local languages. The wide range of information on prices, arrival and other related aspects like grades, standards, packaging, *etc.* is collected and disseminated by networking major agricultural produce markets operating in the country.

E) Akashganga

The initiative facilitates timely collection of milk, proper payments and generates higher income for dairy farmers. The system includes weighment of milk electronically, fat testing, capturing unique ID by the software and printing of pay slip and payment settlement.

F) aAQUA

aAQUA means all most All Question Answered is a multilingual online problem solving system that facilitate farmers getting their queries answered by experts. The reply to the queries raised by the farmers is sent in one to three days depending on the nature of the problem.

G) e-agri Kiosk

An initiative by NABARD and Central Agricultural University. Touch screen kiosk for technology transfer among tribal farmers of Arunachal Pradesh.

Apart from portal, there are so many Mobile App were also developed by various institute/organization with the aim of developing farming community socially, economically and technologically are as follows.

ICAR Mobile Apps	*Organization Developed by*
Agriculture: FEM@Mobile (Farm Extension Manager)	KVK Malappuram; KAU
CAZRI KRISHI	ICAR – Central Arid Zone Research Institute (CAZRI), Jodhpur
Chana Mitra	ICAR – Indian Institute of Pulses Research (IIPR), Kanpur
E-Thilhan	ICAR – Indian Institute of Oilseeds Research (IIOR), Hyderabad
https://play.google.com/store/apps/details?id=in.res.ccari.fertilizercalculator and hl=en Fertilizer Calculator	ICAR – Central Coastal Agricultural Research Institute (CCARI), Goa
GrapesDSS	ICAR; NRCGrapes; IARI; Shivrai Technologies Pvt Ltd.
GypCal –Sodic Soil Reclamation	ICAR – Central Soil Salinity Research Institute, Karnal
CCRI – CITRUS	ICAR – Central Citrus Research Institute, Nagpur; Krish Infotech Systems Pvt. Ltd Education
ICAR – DMAPR	ICAR – Directorate of Medicinal and Aromatic Plants Research, Anand

ICAR Mobile Apps	*Organization Developed by*
ICAR – IIOR	ICAR – Indian Institute of Oilseeds Research (IIOR), Hyderabad
ICAR-IIOR Castor	ICAR – Indian Institute of Oilseeds Research (IIOR), Hyderabad
ICAR – IISR Black pepper	ICAR – Indian Institute of Spices Research (IISR), Calicut; Sourcereflect Technologies LLP
ICAR – MUSHROOM	ICAR – Directorate of Mushroom Research, Solan
ICAR – NIANP FEED CHART	ICAR – National Institute of Animal Nutrition and Physiology (NIANP), Bengsaluru
ICAR – NRCP	ICAR – National Research Centre on Pomegranate, Sholapur
ICAR – NIRJAFT	ICAR – National Institute of Research on Jute and Allied Fibre Technology (NIRJAFT), Kolkata; SPT Global Services
Krishi Vigyan	Krishi Vigyan Kendram, Amadalavalasa, Angrau
KVK Mobile App	ICAR – Indian Agricultural Statistics Research Institute (IASRI), New Delhi
Mango Cultivation IIHR	ICAR – Indian Institute of Horticultural Research (IIHR), Bangalore
Micro Mitra	ICAR – National Bureau of Agriculturally Important Microorganisms (NBAIM), Mau
NRC Banana Info	ICAR-National Reseach Centre for Banana, Trichy
Seed Calculator	ICAR – Central Coastal Agricultural Research Institute (CCARI), Goa
MGR Portal	ICAR – National Bureau of Agriculturally Important Microorganisms (NBAIM), Mau
PaddySeed Xpert	ICAR – National Rice Research Institute (NRRI), Cuttack
Pesticide Calculator	ICAR – Indian Institute of Soybean Research, Indore
Plant Population Calculator	ICAR – Central Coastal Agricultural Research Institute (CCARI), Goa
Pusa Krishi	Mobile Seva: ICAR-Indian Agricultural Research Institute (IARI), New Delhi

Conclusion

As the present situation of human condition is giving an alarming indication that there will be vast demand of food and nutritional crisis in the near future and it has to mitigate by any means otherwise there is something really worst for man-kind is eagerly waiting. Therefore the conservation of resources for human sustenance is must. The conservation of resources is not possible until and unless we do appropriate research on it, educate people on it, create awareness and dissemination of suitable technology, increase of production and productivity, marketing, value addition of products *etc.* Thus from the above context, the role of extension is significant. It can reduce the gap between the researchers and farmers, free from hunger, and nutritional security by bringing the more or less appropriate resource conservation technology in agriculture.

REFERENCES

Anonymous (2013). Climate-Smart Agriculture Sourcebook FAO

Anonymous (2016). Krishi.icar.gov.in, New Delhi

Ensminger D. (1957). A Guide to Community Development. Ministry of Community Development and Cooperation, Govt. of India, New Delhi.

Mathur P.N. (2011). Participatory Technology Development. In future agricultural extension (ed) Kokate et.al. Westville Publishing House, New Delhi.

Mondal, S. (2014). Text book of Agricultural Extension with Global Innovations. Kalyani Publisher New Delhi.

Prasad C., Chaudhary B.N. and Nayar B.B. (1978). First Line Transfer of Technology Project, ICAR, New Delhi

Saravanan, R. (2010). ICTs for Agricultural Extension: Global Experiments, Innovations and Experiences. New India Publishing Agency, New Delhi.

Conservation Agriculture Technologies *Pages* **47–57**
Editor: **Rahul Kumar, Abhinandan Singh and Pankaj Kumar Ojha**
Published by: **BIOTECH BOOKS, NEW DELHI**

4 Impact of Climate Change on Indian Agriculture

***Avinash Goyal*[1]*, Ananta Vashisth*[1] *and Ashish Kumar Singh*[2]**

[1]Division of Agricultural Physics, [2]Division of Nematology, ICAR-Indian Agricultural Research Institute, New Delhi – 110 012

Agriculture sector play acrucial role in growth and development of the Indian economy, but the rise in greenhouse gases creating a pause on this growth and development.The dramatically change in climatic scenario raise a question on food sustainability and quality.Climate change is directly and indirectly affects agricultural crop, livestock, poultry and fisheries in adverse manner. These changes can be seen in various forms such as rise in temperature, variation in rainfall, retreat of glacial and rise of sea level.During the past 115 years since 1901 to 2015 minimum surface air temperature increased 0.85°Cand in the same manner glacier areretreating at a rate of 30 meters every year at Indian subcontinent. The changes in the food habit and metabolic activityof human and animal due to the climate change have physical, financial and ethical implications on the society.Theobjective of this study is to understand the impact of climate change on Indian agriculture andfurther to suggest strategies of adaptation and mitigation to overcome the impact.

1. Introduction

Climate change aggravates the challenges faced by the agriculture sector in all over the world, but developing countries such as India are more susceptible to climate change due to high dependency on agriculture and natural resources. Indian agriculture system provide food to 17.5 per cent of the global population with 2.4 per cent world's land area and 4 per cent water resources (NAAS, 2013). The change in climatic conditions such as increasing earth surface temperature, change in rainfall pattern, occurrence of drought, depletion of groundwater, rising of sea level *etc.* are adding to pressure on the agriculture system that is already in the condition of maelstrom for increasing demand of food and renewable energy. So, self-sufficiency and sustainability in food is an ambiguity. The major cause of climate change has been ascribed to increased level of GHGs due to anthropogenic activities such as fossil fuel burning, deforestation, industrialization,excess use of refrigerators, air conditioner *etc.*The projected change in temperature make a serious problem related to the food security, water scarcity, pollution and human health. The negative impact of climate change has been described by several empirical and modeling approaches on crop yield and livestock at the end of this century (Zhao*et al.* 2017). Average reduction in global yield of cereals would be reduced by about 5 per cent, respectively for each degree of increase in temperature (Zhao*et al.* 2017). A problem related with water are going to be hike with the projected warming and extreme events. Overall, climate change could make it more difficult to grow crops, raise animals, and catch fish in the same ways and same places as we have done in the past.So, keeping the above fact in view it is necessary to discuss the whole issue related with the climate change and its impact on Indian agriculture and ways to overcome it.

2. Statistics of Green House Gases in India

Greenhouse gases are responsible for the global warming resulted into the climate change. The rate of emission of GHGs is drastically increased to 2.2 per cent per year during 2000-2010 compared to 1.3 per cent during the year 1970-2000 (IPCC, 2014).The present concentration of CO_2, CH_4 and N_2O is 410 ppm, 1800 ppb and 320 ppb, respectively. Methane concentration increased about 2.4 times in last 60 years. According to a first biennial update report of India (2015) to the United Nation's Framework Convention on Climate Change (UNFCCC), there is about 9 per cent increase in the GHGs on CO_2 equivalent (eq) within 10 years. According to this inventory agriculture sector emitted 390.1 Tg CO_2 eq. (18.3 per cent of total GHGs emission) in 2010, out of that CH_4 and N_2O are mainly emitted. Methane is mainly emitted from the ruminants due to enteric fermentation and anaerobic rice cultivation. Whereas, N_2O is emitted due to application of nitrogenous fertilizers and alternating wetting and drying of soil. The major sources of emission of GHGs are ruminants, rice cultivation, agricultural soils, residue burning, fertilizers and manure management in the agricultural sector (Figure 4.1).

3. Scenario of Climate Change in India

India is a country with copious numbers of variations in climatic conditions such as temperate in North, tropical in South, arid region in West and humid region in East,although these variation in climatic condition favors the Indian agriculture system due to diversified cropping pattern. But the sudden rise in greenhouse gases (GHGs) after industrialization, affects Indian climate in adverse manner due to wide range of climatic conditions in India. Altogether rise in GHGs leads to various impact on climate.

3.1. Rise in Land and Sea Surface Temperature

The drastic change of climate in the form of the global warming effect is clearly observable after 1960s.The annual maximum and minimum temperature showed increased trend for period 1901 to 2015 (Figure 4.2). During that period the change in minimum surface air temperature is +0.85°C and +0.68°C for meansurface air temperature (IPCC, 2014). According to the IMD report 2013, the Southern states of

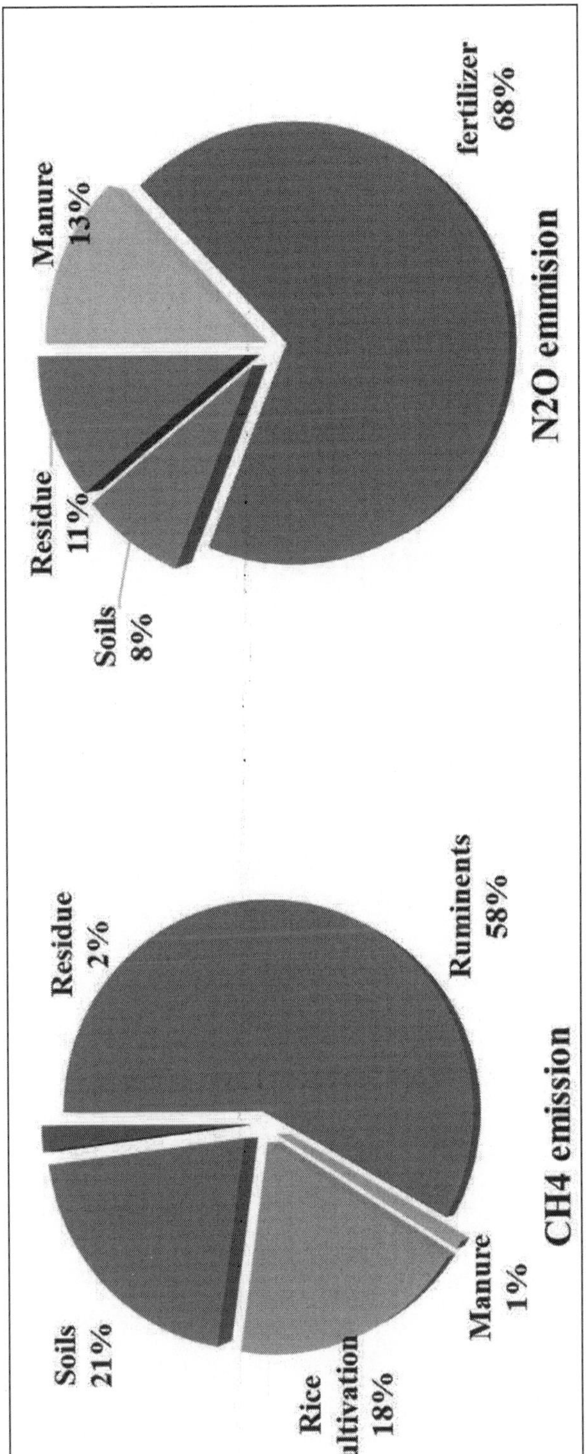

Figure 4.1: Relative Contribution of Various Sources of GHGs Emission from Agriculture (IPCC, 2014).

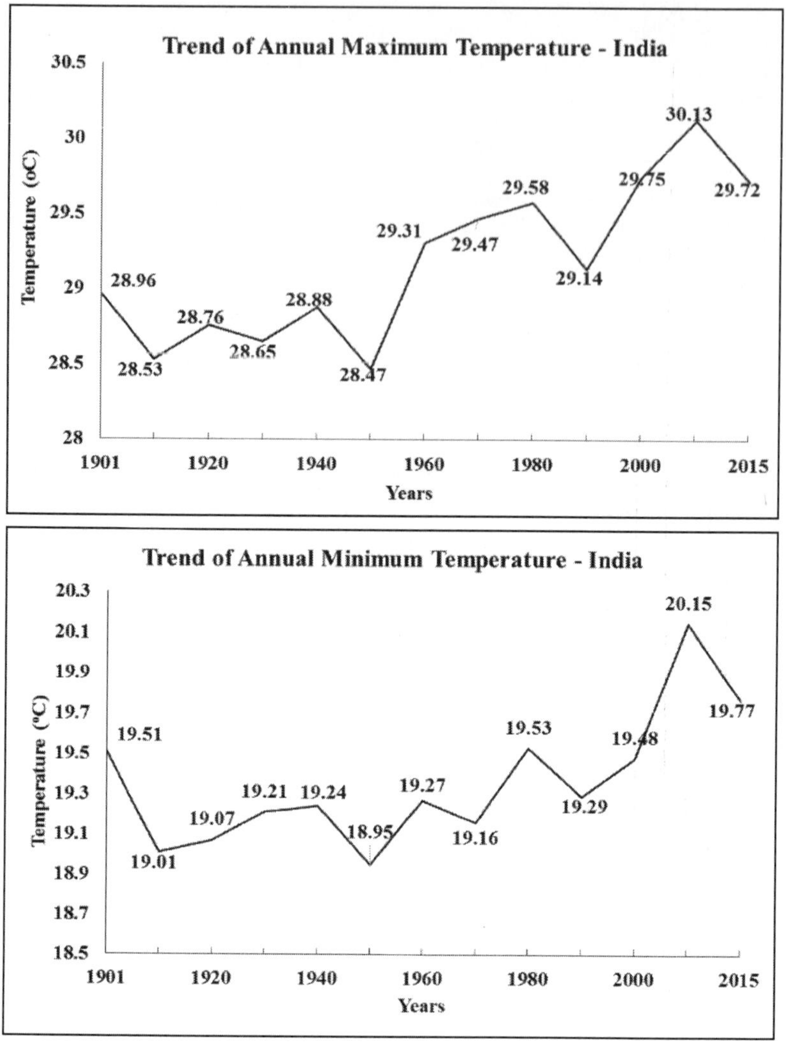

Figure 4.2: Annual Mean Maximum and Minimum Temperature over India (*Source*: MSPI, 2015).

Indian subcontinent having increasing trend in mean maximum temperature,while northern states are having mixed trends with time and place. Recently, according to fifth assessment report of IPCC (2014) the global surface temperature will be exceed 1.5°C at the end of the 21st century.The period and occurrence of extreme events like heat wave have increased over Northwestern region and eastern coast of India. Mohanty*et al.* (2017) analyzed the long term data set and concluded that sea surface temperature of east coast increased by 0.45°C while, 0.67°C along the west coast during last 135 years (1880–2014). The SST anomaly, ranged from –0.19 to +0.32°C and +0.31 to 0.43°C for East and West coast respectively.

3.2 Variation in Rainfall

The tropics are more dependent on agriculture as third forth of the world population lives in tropics.India is a tropical country that have monsoon type of climate. The Indian agriculture is highly sensitive to monsoon variability as 65 per cent of the cropped area is rain-fed.Lots of variability in summer and winter annual rainfall pattern during 1950-51 to 2010 (Figure 4.3) in India rename the Indian agricultureas the *"Gamble of Monsoon"*. India received 75 to 80 per cent rainfall by South-West monsoon (summer monsoon) and 15 to 20 per cent by North-East monsoon (winter monsoon).Climate change start altering these pattern of rain fall where winter monsoon showed the increasing trend in the subdivision of Maharashtra, Telangana, Andhra Pradesh, Northern Karnataka, central part of India and some part of Northwestern India, while summer monsoon faced decreasing trend in the subdivision of Northeast India, Konkan and Goa region. Apart from this summer rainfall showed increasing trend over Rajasthan whilewinter monsoon faced decreasing trend over Tamil Nadu.There are reduction in the number of rainy days during summer monsoon and increased frequency of floods.It is noticed that a poor monsoon decrease agricultural productivity and resulted in inflation although duration and distribution of rainfall is the key factor for the agriculture productivity. Lobell and Burke (2008) also report that change in the rainfall during the growth period by one standard deviation associated with 10 per cent change in production.

3.3. Retreat of Himalayan Glaciers

A glacier is a compact and recrystallized volume of ice and one of the most important source of freshwater. In Indian subcontinent glaciers are found at Himalayas, Kashmir, Uttrakhand, Himachal Pradesh and Northeastern region. The Himalayas is one of the youngest and most rugged mountain ranges.The large geographic extent of the Himalaya, with intricate and extreme topography along with variable climatic conditions, results in an inhomogeneous set of glacial retreats (Dobhal*et al.* 2013). According to Kumar *et al.* (2017),Bangni and Dunagiri glaciers (Uttrakhand) retreated about 2080 ± 162 m and ~484 ± 38 m, respectively between the period of 1962-2013 and the average rate of retreating of these areas was about 41 ± 3.2 m per year and ~9 ± 0.6 m year, respectively which is a clear indication of climate change. He also noticed that Bangni and Dunagiri glaciers lost their height about 17 per cent and 11 per cent, respectively during last 50 year.

3.4. Increase in Sea Level

Melting of glacial leads to the rise in sea level over the past few decades as a consequences of climate change. Further rise in the sea level leads to the decrease in land availability for human living and agricultural purposes.According to the fifth assessment report of IPCC (2014),mean sealevel showed rise trend globally with 1.8 mm per year over the 1961–2003 period and 1.7 mm per year over the 1901–2010 period. Unnikrishnan*et al.* (2015) analyzed the past 55 year data of Indian sea and concluded that the sea level along Bay of Bengal is rising at a faster rate than that of Arabian sea. He also noticed that the net sea level has increased by 1.09 to 1.75 mm/year over Indian sea.

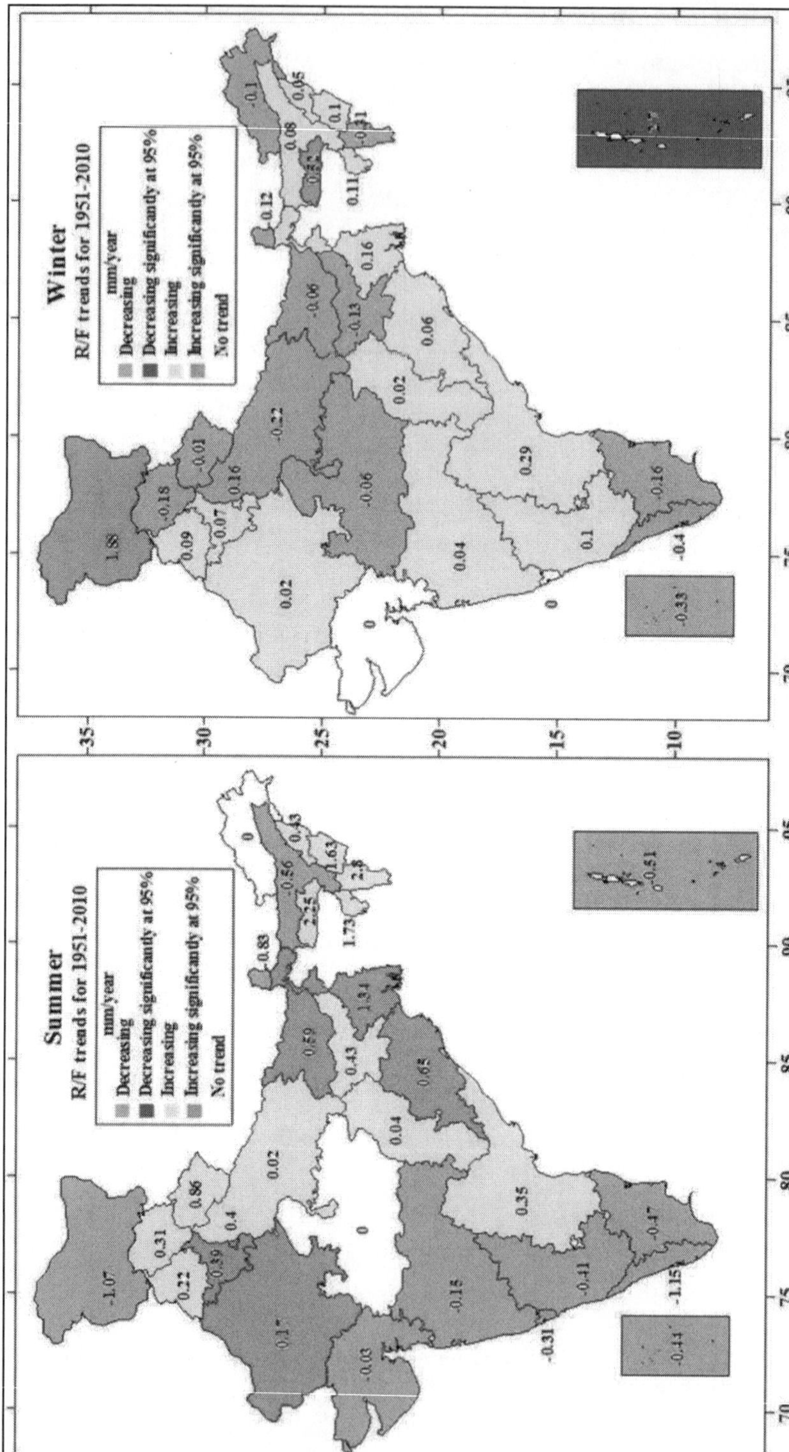

Figure 4.3: Trend of Summer and Winter Rainfall at State Level (*Source*: IMD, 2013).

4. Impact of Climate Change on Agriculture and Allied Sectors

The impact of climate change on Indian agriculture mainly put emphasis at regional level and added to the enormity of India's food security challenges.

4.1. Impact of Climate Change on Agriculture Crops

Climate change is not a single phenomenon that affects agriculture productivity but it is one of the major drivers that is responsible for yield.Indian agriculture is in a peculiar situation of growth with vulnerability due to shifting of crop sown time, sown area, cropping pattern and length of growing seasons.The best example of shifting of cropping pattern can be seen at the change of rice-wheat cropping system into maize -wheat at Punjab and Bihar. Another example is, shifting of cropping area for apple cultivation to higher elevation due to inadequate chilling in Himachal Pradesh.The developing country such as India shows drastically lower yield than those of developed countries as North America, East Asia and the Pacific and the Euro area (Chakrabarty, M., 2016).The wheat, rice, maize and soybean are major cereals provides two-thirds of human caloric intake in India. Zhao *et al.* (2017) investigatedlong term data (about 100 years) and observed the impact of temperature on yields for major cereals and concluded that loss in the yieldfor each degree increase in mean temperature is largest for wheat (–9.1 ± 5.4 per cent followed by rice –6.6 ± 3.8 per cent and maize –5.2 ± 4.5 per cent. Reduction in the yields of wheat, rice, maize and soybean by 6.0. 3.2, 7.4, and 3.1 per cent, respectively on the global level.The tea crop in mid hill regions also showed a decrease in yield with increasing temperature and decreasing rainfall.Potato yield reduction in Haryana, Punjab and western and central Uttar Pradesh by 3.46 to 7.11 per cent in, but in rest of India particularly West Bengal and plateau region potato production may decline by 4 to 16 per cent (Venkateswarlu, B., 2011).

4.2. Impact of Climate Change on Livestock

India is leading in livestock population over the globe out of that 1stin cattle and buffalo, 2nd in goat population and 3rd in sheep (GOI, 2014). The climate affects the livestock both directly (diseases, reproduction, biodiversity, animal growth and milk production) and indirectly such as changes in production and quality of feed crop and forage, water availability (Rojas-Downing *et al.*, 2017).Increase in temperature, variability in rainfall and CO_2 concentration have an adverse effect on livestock. The basic physiological mechanisms of ruminants are negatively affected with increase the air temperature that increased risk of different metabolic disorders, health problems and reproductive problems (Soriani *et al.*, 2013).The effect of temperature and relative humidity on comfort level of livestock has been computed through formulating temperature humidity index (THI).THI was positively related to breathing rate and negatively related to milk yield (Soriani *et al.*, 2013).Heat stress is responsible for 10 to 30 per cent reduction in milk yield in first lactation and 5-20 per cent in second and third lactation. The stress level is increased with THI such as mild (70-75), moderate (75-80), less severe (80-85) and more severe >85. Changes in the food is also modified the metabolic activity of livestock. Less roughages consumed by the ruminant is able to change rumen microbial population and pH from 5.82 to 6.03 decreasing rumen motility and rumination (Soriani *et al.*, 2013).

4.3. Impact of Climate Change on Poultry

Change in the food habits leads to increase in demands ofeggs and chicken is increasing continuously in India.Poultry survives in narrow temperature range because birds are more sensitive to temperature change due to lack of the sweat glands. The rate of mortality was shown a positive correlation with the ambient temperature in the case of poultry. The body weight of broiler is negatively related to the temperature. Olanrewaju*et al.* (2010) observed that body weight of boiler decreased from 4.08 to 3.35 kg relatively small changes in the ambient temperature (18 to 24°C). Overall mortality (per cent) was increased with rise in the ambient temperature of broiler, layer and native chickens during 2004 to 2009 at Project Directorate of Poultry, Hyderabad (Rao, 2017) The broiler type chickens are ten time more sensitive to heat compared to layer type chickens (Figure 4.4). Star *et al.* (2009) recorded a significant reduction in feed conversion, egg production and egg weight of 31.6, 36.4 and 3.41 per cent, respectively to heated environment.Heat stress is also responsible in reduction of egg weight (–3.24 per cent), egg shell thickness (–1.2 per cent), eggshell weight (–9.93 per cent), and eggshell per cent (–0.66 per cent).

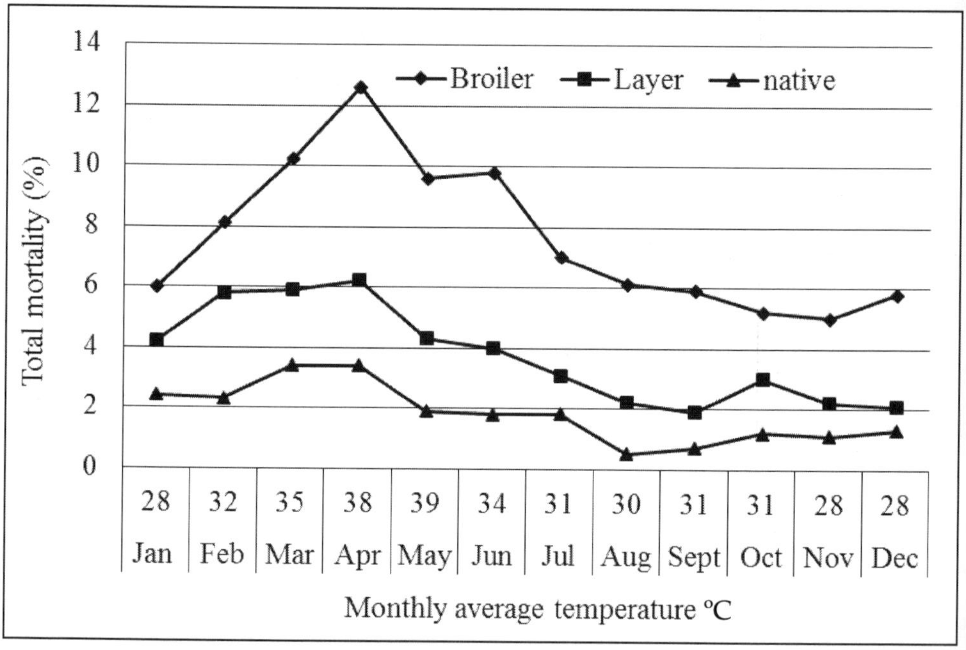

Figure 4.4: Effect Ambient Temperature on the Mortility of Broiler, Layer and Native Chicken (Rao, 2017).

4.4. Impact of Climate Change on Fisheries

Marine fisheries is important to nutritional security, income and employment generation and foreign exchange earnings for the country. In 2014, the marine fish landings were 3.59 million ton, valued at US$5.6 billion at landing center price (CMFRI, 2014). There are already many problems faced by fisheries including overfishing and water pollution and along with these climate change making the situation more worse. In particular, sea surface temperature (SST) changes found positive correlation the horizontal and vertical extension of distributional boundary for catches. Fisheries respond to climate change bypeak spawning shifting towards cooler months,faster growth rate,reduction in mean length in fishery due to seawater warming andreduction in number of eggsdue toadvanced maturity (Mohanty *et al.*, 2017).

5. Adaptation and Mitigation Strategies of Climate Change in Agriculture

It is undoubtable thing that if we do not take instant action against climate change than this problem is going to be much more than just an economic threat. It is a wrong conception that adaptation strategies are sufficient for climate change without considering mitigation plans. Adaptation can reduce the risks of climate change, but there are limits to its effectiveness, especially with greater magnitudes and rates of climate change. So, It is important to focused more on mitigation strategies for climate change.

5.1. Adaptation Strategies

- ☆ Development of resistant varieties for drought, frost, flood, high CO_2 concentration and salt.
- ☆ Changing the location and timing of cropping activities
- ☆ Development of short duration varieties
- ☆ Diversified agricultural system such as livestock raising and fish culture in rice field.
- ☆ Seasonal weather forecasting used for farm management operations to reduce production risk.
- ☆ Practicing agroforestry
- ☆ Agriculture-insurance for compensate the losses in crop yield due to climate variability
- ☆ Efficient use of fertilizer such as optimum dose, deep placement of nitrogen fertilizers, use of neem coated urea, band placement of potash fertilizer *etc.*

5.2. Mitigation Strategies

- ☆ Conservation agriculture
- ☆ Reduction of GHGs
- ☆ Rainwater conservation and harvesting
- ☆ Soil carbon sequestration
- ☆ Anaerobic cultivation of rice
- ☆ Site specific nutrient management
- ☆ Afforestation

- ☆ Biomass energy and waste recycling
- ☆ Encourage organic farming and biomass fuel
- ☆ Development of warehouses to ensure local food security

6. Conclusion

Rise of industrialization around the globe and India lead to rise of GHGs and ultimately the climate change. Industrial emissions of carbon dioxide and other greenhouse gases will cause global warming, with an increase in the earth's temperature of 2.8°C to 5.2°C over the next 50 to 100 years. Although India is start facing the ill effect of climate change but still coming to any concrete conclusion or any strategies. Without, a certain assessment of the impacts and vulnerabilities of agriculture sector it is hard to design any policy. So it is necessary to use multi-pronged strategy that creates wide range of adaptation of existing technologies to control the climate change. Policy incentives will also play crucial role in adoption of climate ready technologies in agriculture too as in other sectors.

REFERENCES

Chakrabarty, M. 2016. Climate change and food security in India. ORF issue briefs and special reports.

CMFRI 2014. Annual Report. Central Marine Fisheries Research Institute, Kochi, India.

Dobhal, D. P., Mehta, M., Srivastava, D. 2013. Influence of debris cover on terminus retreat and mass changes of Chorabari Glacier, Garhwal region, central Himalaya, India. *Journal of Glaciology*, **59**(217): 961–971.

Government of India (GOI). 2014. Basic Animal Husbandry Statistics. Ministry ofAgriculture. Department of Animal Husbandry, Dairying and Fisheries. New Delhi.

IMD Meteorological Monograph No. ESSO/IMD/EMRC/02/2013. State level climate change trends in India.

India First Biennial Update Report to theUnited Nations Framework Conventionon Climate Change, 2015.

IPCC, 2014: Climate Change: Synthesis Report. Contribution of Working Groups I, II and III to the Fifth Assessment Report of the Intergovernmental Panel on Climate Change [Core Writing Team, R.K. Pachauri and L.A. Meyer (eds.)]. IPCC, Geneva, Switzerland, 151 pp.

Kumar, V., Mehta, M., Mishra, A., Trivedi, A. 2017. Temporal fluctuations and frontal area change of Bangni and Dunagiri glaciers from 1962 to 2013, Dhauliganga Basin, central Himalaya, India. *Geomorphology*, **284**: 88-98.

Lobell, D. B., Burke M. B. 2008. Why are agricultural impacts of climate change so uncertain? The importance of temperature relative to precipitation. *Environmental Research Letters*,**3**: 1–8.

Ministry of Statistics and Programme Implementation. 2015. Statistics Related to Climate Change – India-2015.

Mohanty, B., Vivekanandan, E., Mohanty, S., Mahanty, A., Trivedi, R., Tripathy, M., Sahu, J. 2017.The Impact of Climate Change on Marine and Inland Fisheries and Aquaculture in India. *Climate Change Impacts on Fisheries and Aquaculture: A Global Analysis*, Chapter 17: 569-601.

NAAS2013. Climate resilient agriculture in India. Policy papers No 65, New Delhi.

Olanrewaju, H. A., Purswell, J. L., Collier, S. D.,Branton, S. L. 2010. Effect of ambient temperature and light intensity on physiological reactions of heavy broiler chickens *Poultry Science*, **89**(12): 2668–2677.

Rao., G. S. L.H. V. P. 2017. Climate change: Adaptations of livestock, poultry and fisheries. *Agriculture under Climate Change: Threats, Strategies and Policies*, Allied Publishers Pvt. Ltd.pp- 206-212.

Rojas-Downing, M. M., Nejadhashemi, A. P., Harrigan, T., Woznicki, S. A. 2017. Climate change and livestock: Impacts, adaptation, and mitigation. *Climate Risk Management*, **16**: 145–163.

Soriani, N., Panella, G., Calamari, L. 2013. Rumination time during the summer season and its relationships with metabolic conditions and milk production. *Journal of Dairy Sciences*, **96**(8): 5082–5094.

Star, L., Juul-Madsen, H. R., Decuypere, E., Nieuwland, M. G., Reilingh, D. G., Brand, V. H., Kemp B., Parmentier, H. K. 2009. Effect of early life thermal conditioning and immune challenge on thermo-tolerance and humoral immune competence in adult laying hens. *Poultry Science*, **88**: 2253–2261.

Unnikrishnan, A. S.,Nidheesh, A. G.,Lengaigne, M. 2015. Sea-level-rise trends off the Indian coastsduring the last two decades. *Current Science*, **108** (5): 966-971.

Venkateswarlu, B. 2011.Agriculture and climatechange. Presented at the south asian media briefing workshop on climate change,Delhi.

Zhao, C., Liu, B., Piao, S., Wang, X., Lobell, D. B., Huang, Y., Huang, M., Yao, Y., Bassu, S., Ciais, S., Durand, J. L., Elliott, J., Ewert, F., Janssens, I. A., Li, T., Lin, E., Liu, Q., Martre, P., Müller, V., Peng, S., Peñuelas, J., Ruane, A. C., Wallach, D., Wang,T., Wu, D., Liu, Z., Zhu, Y., Zhu, Z., Asseng, S. 2017.Temperature increase reduces global yields of major crops in four independent estimates. *Proceedings of the National Academy of Sciences*, **114**: 9326–9331.

Conservation Agriculture Technologies *Pages* **59–65**
Editor: **Rahul Kumar, Abhinandan Singh and Pankaj Kumar Ojha**
Published by: **BIOTECH BOOKS, NEW DELHI**

5 Potential of Agricultural Crop Residues in the Production of Mushroom

Rakhi Murmu[1] and Rahul Kumar[2]

[1]Regional Plant Quarantine Station, Salt Lake, Kolkata – 700 097, West Bengal
[2]ICAR-Research Complex for Eastern Region, Patna – 800 014, Bihar

Mushrooms are a group of saprophytic fungi. They are not able to synthesize their own food because of the absence of chlorophyll, and hence are dependent upon the organic matter/substrate for food. Mushrooms have been used by humans for millennia. Its fruiting bodies have been known as food source for thousands of years. Edible mushrooms are well known for its delicacy and are integral part in the cooking of various cultures (Shah *et al.*, 2013). They are included in Basidiomycota and some members in Ascomycota. More than 2000 species of mushrooms are present in nature but around 22 are widely accepted for cultivation, commercial purposes, on ground or wood and utilizing particular environmental and nutritional conditions. Besides their diverse and interesting culinary uses, mushrooms have nutritional and medicinal value.

The use of mushrooms in traditional medicine has been documented throughout Asia, Europe, Africa, and Mesoamerica. Total of more than 130 medicinal functions are thought to be produced by medicinal mushrooms including immunomodulating, antitumor, antioxidant, radical scavenging, cardiovascular, anti-hypercholesterolemia, antifungal, antibacterial, antiviral, anti-parasitic, detoxification, hepato-protective, anti-diabetic, and many other effects. The best known medicinal mushrooms are: *Ganoderma lucidum* (lingzhi or reishi), *Lentinula edodes* (shiitake), *Grifola frondosa* (maitake), *Cordyceps* species (caterpillar mushrooms), *Trametes versicolor* (turkey tail), *Flammulina velutipes* (winter mushroom or enokitake), *Pleurotus* species (oyster mushrooms), *Agaricus brasiliensis* (royal sun mushroom), *Hericium erinaceus* (lion's mane), *Hypsizygus marmoreus* (beech mushroom), *Tremella mesenterica* (yellow brain mushroom), *T. fuciformis* (silver ear mushroom), *Phellinus linteus* (black hoof mushroom), and *Inonotus obliquus* (chaga). Mushrooms are good sources of protein, vitamins: A, B1, B2, B3, C, D2, phosphorus, iron, calcium, and other minerals.

Table 5.1: Some of the Commercially Cultivated Varieties of Mushrooms

Common Name	*Scientific Name*
White button mushroom	*Agaricus bisporus*
Oyster mushroom	*Pleurotus ostreatus*
Paddy straw mushroom	*Volvariella volvacea*
Milky mushroom	*Calocybe indica*
Shiitake mushroom	*Lentinus edodes*
Wood ear mushroom	*Auricularia auricular*
Enoki mushroom	*Flammulina velutipes*
Bear's head mushroom	*Hericium erinaceus*
Silver ear mushroom	*Tremella fuciformis*
Lingzhi mushroom	*Ganoderma lucidum*

Mushrooms can be grown by anyone, from backyard gardeners to large-scale corporations. They do not require arable land, large amounts of water, or special light. Mushrooms do not compete with other plants, and even enhance the growth of many plants. As natural resources are limited, sustainability becomes an important concept in maintaining the human population, health and environment. Mushroom cultivation is a direct utilization of their ecological role in the bioconversion of solid wastes generated from agriculture industry and into edible biomass, which could also be regarded as a functional food or as a source of drugs and pharmaceuticals. Mushrooms can be grown indoors as well as outdoors.

Organic solid wastes are a kind of biomass, which are generated annually through the activities of the agricultural, forest and food processing industries. They have insignificant or less commercial value and certainly no food

Figure 5.1: *Agaricus bisporus.*

Figure 5.2: *Pleurotus ostreatus.*

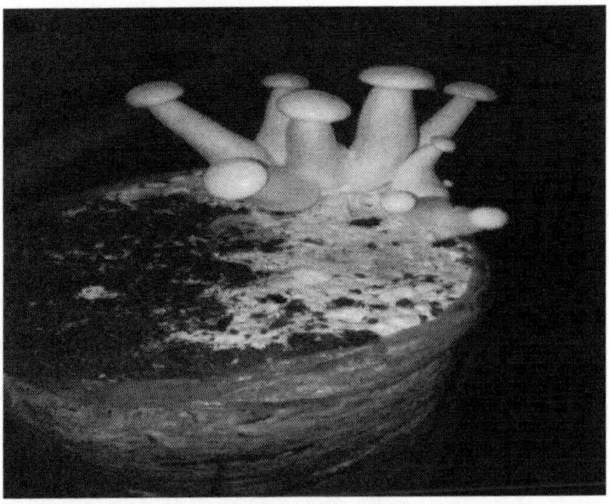

Figure 5.3: *Calocybe indica.*

value. These wastes can be converted into usable resources through proper management, with their utilization leading to reduced environmental pollution and further economic growth. Most agricultural residues are rich in lignocellulosic compounds whose handling and disposal are often problematic, due to their chemical structure and decomposition properties. As they are produced in large quantities, their post-harvest treatment is mainly accomplished through burning or incorporation into the soil. Many low-grade agricultural waste products including the stalks of agricultural produce, corn cobs, cotton shells, sugar cane segments, sugar beets, industry waste from cotton mills, slaughterhouses, meat processing plants, paper factories, sewage from livestock feedlots, *etc.* can be used as positive alternative substrates for mushroom cultivation (Zervakis and Balis, 1992; Ragunathan *et al.*, 1996; Zervakis *et al.*, 1996; Thomas *et al.*, 1998; Philippoussis *et al.*, 2000). Cellulose and some nutrients must be added to the waste materials for faster rapid growth. According to the Food and Agriculture Organization, about 20 million tones of usable agricultural wastes are discarded every year. An English researcher found that developed countries burn about 60 per cent of usable waste products each year. For East Asian farmers, mushroom production is an important part of their sustainable agriculture system. Recycling waste products form mushroom cultivation is one of the best method of conservation of our natural resources.

Mushroom cultivation is an economically important biotechnological industry that has been distinguishly expanded worldwide in the last few decades. Its cultivation mainly depends on the agricultural crop residues. These agricultural residues are abundantly available in our country. It is primarily used as cattle feed and remaining part is burnt or spread in the field (Ram and Kumar, 2010). Agro-wastes contain about 80 per cent cellulose, hemi-cellulose, and lignin which are not easily degradable, but most of the edible fungi posses enzyme system which degrade these components. Mushrooms, with other fungi, are presently only organisms that can synthesize and excrete the relevant hydrolytic and oxidative enzymes that enable them to degrade complex organic substrates into soluble substances, which can then be absorbed by the mushrooms for their nutrients (Chang and Miles, 2004).

The potential of bioconversion of lignocellulosic wastes into value-added products is emphasized in recent studies (Philippoussis and Zervakis, 2000; Poppe, 2000). Edible mushroom fungi possess the appropriate enzymatic mechanisms for the transformation of complex organic macromolecules into simple compounds. This bioconversion has been exploited as the means for the biodegradation of a wide range of plant litter due to their particular ability for selective delignification (Mayson and Verachtert, 1991; Martinez *et al.*, 1994). This conversion of useless and potentially hazardous agricultural wastes into food of high organoleptic properties and nutritive value (Rambelli, 1985), or good quality fodder by enriching the waste protein content and by improving their digestibility (Martinez, *et al.*, 1991; Tripothi and Yadar, 1992), or production of plant fertilizers and soil conditioners beneficial to overall soil fertility and stability, as well as vegetable yield. In addition, the production of mushroom species benefits the social development and promotion of employment in rural areas. Different species of mushrooms have different abilities to utilize the substrates. This depends on the particular enzymes secreted by the individual mushroom. *Lentinula edodes*, cultivated on sawdust or wood, highly lignified substrates, produces two extracellular enzymes (manganese perxidase and laccase), which have been associated with lignin depolymerization. On contrary, *Volvariella volvacea* prefers high-cellulose, low lignin-containing substrates, such as paddy straw and cotton wastes, which have relatively low-lignin content, and it produces a family of cellulolytic enzymes including at least five endoglucanases, five cellobihydralases, and two ß-glucosidases, but none of the recognized lignin-degrading enzymes. *Pleurotus pulmonarius* var. *stechangii* (an oyster mushroom) is the most adaptable of the three species and can be grown on a wide range of agricultural waste materials of different composition in terms of polysaccharide/lignin ratio, because it can excrete both kinds of cellulose and lignin-degrading enzymes. Lignocellulosic compounds are complex and insoluble. They can be treated by various chemical methods, for example, with calcium chloride and dilute hydrochloric acid to increase the digestibility and nutritional qualities, and even to form sugars to serve as carbon sources. However, these chemical methods are tedious and costly. Furthermore, treatments to eliminate adverse side effects of the chemicals are also very complex. In contrast, mushroom cultivation techniques have become significantly important in recent years to improve nutritional quality and to upgrade the economic value of the solid organic wastes.

Thus, mushroom cultivation can avoid environmental pollution by recycling agricultural wastes and also convert straw into easily digestible animal feed. About 385 million tonnes of agricultural crop wastes are available annually in India about half of this residue remains unused. If even one per cent of these residues are used for mushroom production, India will become a major mushroom producing country in the world. It is highly profitable business for gaining employment for the poor people and is labour-intensive. The cultivation of mushroom can be an eco-friendly alternative for recycling agro-waste, providing better nutrition for the vast vegetarian population, employment generation and a good source of income.

One of the primary roles of mushrooms in the ecosystem is decomposition, which is performed by the mycelium. Mushroom mycelium can produce a group of complex extracellular enzymes, which degrades and utilizes the lignocellulosic wastes to reduce pollution. Mushroom mycelia can also play a significant role in the restoration of damaged environments. Stamets (2005) has coined a term, mycorestoration, which can be performed in four different ways: mycofiltration (using mycelia to filter toxic wastes and microorganisms from water in soil film in the air), mycoforestry (using mycelia to restore forests), mycoremediation (a form of bioremediation using mycelia to decontaminate the area), and mycopesticides (using mycelia to control insect pests). These methods represent the potential to create a clean ecosystem, where no damage will be left after fungal implementation, even if there are some toxic wastes.

Biosorption is the process in which microorganisms are used to remove and recover heavy metals from aqueous solution, an outcome of physico-chemical interactions between metal ions and cellular compounds of species, mainly comprising chemical groups such as carboxyl, carbonyl, amino amido, sulphonate, phosphate, and hydroxyl (Kapoor and Viraraghavam, 1998). Mushrooms are being looked at as an environment cleaning tool, which can be used to sequester heavy metals released into natural water bodies as industrial effluents (Nilanjana, 2005). Mushrooms can uptake heavy metals from substrate through the spacious mycelium by biosorption. This phenomenon has also been reported in many microorganisms and plant materials (Nilanjana *et al.*, 2008; Innocent *et al.*, 2009).

Spent mushroom substrate is the remainant of compost from which mushrooms have been produced. It is bulky solid waste generated from the mushroom industry, which can be utilized profitably in different ways such as soil fertilizer and a prospective bioremediating agent (Masaphy *et al.*, 1987). The lignocellulosic substrate, used for mushroom production after harvesting the mushrooms, is used for soil conditioning or organic fertilizer as compost. This compost, besides being rich in nitrogenous material, contains partly degraded lignocellulosic components that, when combined with pure animal dung or human excrement in a biogas digester, yields not only biogas but is also a good quality organic nitrogenous fertilizer in the form of sludge. The sludge from the biogas plant as a nitrogenous fertilizer is far more beneficial than the compost from which it has been derived. These wastes can be converted into valuable resources through proper management, with their utilization leading to reduced environmental pollution and further economic growth. When carelessly disposed of in the surrounding environment by dumping or burning, these wastes are bound to lead to environmental pollution and, consequently, to health hazards (Stamets, 2005).

The pleasing flavor, high protein level and medicinal values, mushrooms clearly represent one of the world's greatest untapped resources of nutritious and palatable food for our current generation and for future generations to come. The mushroom life cycle is not only shorter than many crops, but the yield is very high. Not only do mushrooms provide food, but mushroom waste can be recycled into fertilizers and additives that improve tree plantation soil conditions. It is also evident that a lot of mushrooms are available having the capacity of effective biosorption of heavy metals which can be taken as a low cost technology to solve the problem of heavy metal pollution. People often find mushroom cultivation is not as easy as they thought it would be. Mushroom growers need both experience and education. Much of the necessary knowledge must be acquired through practical experience, but understanding the principles of mushroom cultivation demystifies the process, allowing the grower to successfully adapt and develop cultivation methods. Mushrooms are environment friendly. Furthermore, their production and distribution can serve as agents for promoting equitable economic growth in society.

REFERENCES

Chang, S. T. and Miles, P. G. (2004). Mushroom: Cultivation, nutritional value, medicinal effect, and environmental impact (2d ed.). Boca Raton, FL: CRC Press.

Innocent, O., Emmanuel, A. and Thomas, A. (2009). Biosorption of heavy metal ions from aqueous solutions using a biomaterial. *Leonardo Journal of Science*. 48: 58-65.

Kappor, A. and Viraraghavan, T. (1998). Biosorption of heavy metals of *Aspergillus niger*, effect of pretreatment. *Bioresource Technology*. 63: 109.

Martinez, A.T., Gonzalez, A. E., Valmaseda, M., Dale, B. E., Lambregts, M. J. and Haw, J. F. (1991). Solid-state NMR studies of lignin and plant polysaccharides degradation by fungi. *Holzforschung*. 45: 49–54.

Martinez, A. T., Camarero, S., Guillen, F., Gutierrez, A., Munoz, C., Varela, E., Martinez, M. J., Barrasa, J. M., Ruel, K. and Pelayo, M. (1994). Progress in biopulping of non-woody materials: chemical, enzymatic and ultrastructural aspects of wheat-straw delignification with ligninolytic fungi from the genus *Pleurotus*. *FEMS Microbiology Reviews*.13: 265–274.

Masaphy, S., Levanon, D., Danai, O. and Henis, Y. (1987). Nutritional supplementation to the casing soil. Ecological aspects and mushroom production mushroom Science (Part II 0). In Proccedings of 12th International Congress on Science and cultivation of edible fungi (Braunschweig, Germany), 417-426.

Mayson, E. and Verachtert, H. (1991). Growth of higher fungi on wheat straw and their impact on the digestibility of the substrate. *Applied Microbiology and Biotechnology*.36: 421–424.

Nilanjana, D. (2005). Heavy metals biosorption by mushrooms. *Natural Product Radiance*. 4(6): 454–459.

Nilanjana, D., Karthika, P., Vimala, R. and Vinodhini, V. (2008).Use of natural products as biosorbent of heavy metals – An overview. *Natural Product Radiance*. 7(2): 133–138.

Philippoussis, A. and Zervakis, G. (2000). Cultivation of edible mushrooms in Greece: presentation of the current status and analysis of future trends. In Proceedings of the 15th International Congress on the Science and Cultivation of Edible Fungi, ed. Van Griensven, L.J.L.D., pp. 843–848. Rottterdam: Balkema. ISBN 90–5809–1457.

Philippoussis, A., Zervakis, G., Diamantopoulou, P. and Ioannidou, S. (2000). Potential for the cultivation of exotic mushroom species by exploitation of Mediterranean agricultural wastes. In Proceedings of the 15th International Congress on the Science and Cultivation of Edible Fungi, ed. Van Griensven, L.J.L.D., pp. 523–530. Rottterdam: Balkema. ISBN 90–5809–1457.

Poppe, J. (2000). Use of agricultural waste materials in the cultivation of mushrooms. In Proceedings of the 15th International Congress on the Science and Cultivation of Edible Fungi, ed. Van Griensven, L.J.L.D., pp. 3–23. Rottterdam: Balkema. ISBN 90-5809-1449.

Ragunathan, R., Gurusamy, R., Palaniswamy, M. and Swaminathan, K. (1996). Cultivation of *Pleurotus* spp. on various agro-residues. *Food Chemistry*. 55: 139-144.

Ram, R. C. and Kumar, S. (2010). Agricultural wastes used as casing mixtures for production of button mushroom. *Indian Journal of Scientific Research*. 1 (1): 21-25.

Rambelli, A. (1985). Manual on mushroom cultivation. FAO plant production and protection paper 43. FAO, Rome. ISBN 92-5-101324-1.

Shah, S., Nasreen, S. and Kousar, S. (2013). Efficacy of fungicides against *Trichoderma* spp. causing green mold disease in oyster mushroom (*Pleurotus sajor-caju*). *Research Journal of Microbiology*. 8(1): 13-24.

Stamets, P. (2005). Mycelium running: How mushrooms can help save the world. Berkeley, CA: Ten Speed Press.

Thomas, G.V., Prabhu, S. R., Reeny, M. Z. and Bopaiah, B. M. (1998). Evaluation of lignocellulosic biomass from coconut palm as substrate for cultivation of *Pleurotus sajor-caju* (Fr.) Singer. *World Journal of Microbiology and Biotechnology*. 14: 879-882.

Tripothi, J. P. and Yadar, J. S. (1992). Optimization of solid substrate fermentation of wheat straw into animal feed by *Pleurotus ostreatus* — a pilot effort. *Animal Feed Science and Technology*. 37: 59-72.

Zervakis, G. and Balis, C. (1992.) Comparative study on the cultural characters of *Pleurotus* species under the influence of different substrates and fruiting temperatures. *Micologia Neotropical Aplicada*. 5: 39-47.

Zervakis, G., Yiatras, P. and Balis, C. (1996). Edible mushrooms from olive mill wastes. *International Biodeterioration and Biodegradation*. 38: 237-243.

Conservation Agriculture Technologies *Pages* **67–75**
Editor: **Rahul Kumar, Abhinandan Singh and Pankaj Kumar Ojha**
Published by: **BIOTECH BOOKS, NEW DELHI**

6 Influence of Crop Residue Management on Soil Health

Pooja Kumari and D.K. Roy

Department of Agronomy, Dr. Rajendra Prasad Central Agricultural University, PUSA, Samastipur – 848 125, Bihar

Crop residues are the plantspart left in the field after crops have been harvested and threshed. The recycling of crop residues has the advantage of converting the surplus farm waste into useful product for meeting nutrient requirement of succeeding crops. Crop residues are a good source of organic C for soil microbes and also contribute to plant nutrients.Crop residue in fields can be considered key factor in promoting physical, chemical, and biological attributes of soil health in agricultural systems. It is clear that the main obstacle to improving soils and other ecosystems through conservation agriculture in these situations is the lack of residues produced and the competition for alternate, higher value use of residues. However, due to multiple other uses, farmers are faced problem in managing crop residues. The benefits of residue retention are regionally variable and depend on both agro-climatic and socioeconomic factors.Crop residues contain significant quantities of plant nutrients and their judicious application will have positive effect on nutrient availability in soil. Positive effects of retaining crop residues on soil quality, soil organic matter and carbon storage, soil moisture retention, enhanced nutrient cycling, and decreased soil loss, among other environmental and soil health benefits. In addition to the role of the organic matter from decomposition of crop residue in increasing the concentration of micronutrients in soil solution is very important.The decreased soil temperature, nitrogen immobilization, and waterlogging are some of the limitations of crop residue retention on soil surface.In developing countries, the key in promoting physical, chemical, and biological attributes of soil health in agricultural systems is mainly through the retention of crop residue in fields in view of scarcity of alternative source of organic amendments.

Introduction

India is an agricultural based country. Majority of land is used for farming and a wide range of crops are cultivatedon the different agro-ecological regions of the country. Crop residue is the vegetative crop material left on the field after a crop is harvested, pruned or processed. The Ministry of New and Renewable Energy

(MNRE, 2009), Govt. of India has estimated that about 500 Mt of crop residues are generated every year.The generation of crop residues is highest in Uttar Pradesh (60 Mt) followed by Punjab (51 Mt) and Maharashtra (46 Mt). Among different crops, cereals generate maximum residues (352 Mt), followed by fibres (66 Mt), oilseeds (29 Mt), pulses (13 Mt) and sugarcane (12 Mt). The cereal crops (rice, wheat, maize, millets) contribute 70 per cent while rice crop alone contributes 34 per cent to the crop residues. Wheat ranks second with 22 per cent of the crop residues The Rice-Wheat cropping system produced nearly one-fourth of the total Crop Residues in India (Singh and Sindhu,2014), whereas fibre crops contribute 13 per cent to the crop residues generated from all crops. Among fibres, cotton generates maximum (53 Mt) with 11 per cent of crop residues (MNRE, 2009). Crop residues are mainly used for Livestock feeding, Mulching, Manure production, thatching for rural homes and fuel for domestic and industrial use. Crop residues are good sources of plant nutrients and are important components for the stability of agricultural ecosystems. In areas where mechanical harvesting is practiced, a large quantity of crop residues are left in the field, which can be recycled for nutrient supply. Incorporation of crop residues into the soil as a source of nutrients and organic matter, which increases soil quality (Choudhury *et al.*, 2014; Soon and Lupwayi, 2012).A large portion of the crop residues is not utilized and left in the fields. The disposal of such a large amount of crop residues is a major challenge. Farmers opt for burning because it is a quick and easy way to manage the large quantities of crop residues and prepare the field for the next crop well in time. To manage the residues in a productive and profitable manner, conservation agriculture (CA) offers a good promise. With the adoption of conservation agriculture-based technologies like reduced or zero tillage and mulching, these residues can be used for improving soil health, increasing crop productivity, reducing pollution and enhancing sustainability and resilience of agriculture.Crop residues are generally thought to enhance and protect soil health. Effects of crop residues left on the soil surface on soil functions *i.e.*, protection from erosive forces, increased or maintained soil organic matter, additions to the available pool of soil nutrients, increased biological activity which improved soil structure and ultimately improved crop yields (Hargrove, 1991).In addition to the main nutrients (N, P and K), crop residues contain also substantial amounts of secondary nutrients and micronutrients, then returning back these residues into the soil may be one of the best alternative practices for improving the physical, chemical and biological properties of the poor soils (Hiel *et al.*, 2016).Thus this chapter focus on the effect of crop residue management on soil quality and health in order to examining physical, chemical and biological properties of agricultural soils in India.

Effect of Crop Residue on Soil Health and Quality

Soil health is defined as "the capacity of soil to function as a living system" (FAO, 2011), while soil quality is its "fitness for use" (Larson and Pierce, 1994). In an agricultural context, high soil quality means a highly productive soil with low levels of degradation (Fuentes *et al.*, 2009) So, Soil quality for sustainable crop production is related to soil health. (Turmel *et al.*, 2015). Crop residue gives organic matter to the soil where it is retained through a combination of physical, chemical, and biological activities that interact and affect soil quality in agricultural soils

(Wilhelm *et al.*, 1986; Wilhelm *et al.*, 2007). The influence of residue management on some important chemical properties (soil organic carbon, soil pH nutrient availability and cation exchange capacity), on physical properties (soil structure, runoff, soil loss, soil compaction, soil temperature, and moisturecontent), and on biological properties (soil biodiversity and soil microbial biomass) are discussed below.

Effect of Crop Residue on Soil Physical Properties

1. Soil Structure

Retention of crop residue can improve soil structure through various mechanisms: (1) increasing soil aggregation due to adding organic matter to the top soil, (2) protecting soil aggregates from impact of raindrop, and (3) protecting soil from compaction caused by raindrop impact (Verhulst *et al.*, 2010a). According to Panachuki *et al.* (2011), residue cover is the most important factor in reducing the effect of rainfall and thus protecting soil aggregates from breaking down.

2. Surface Runoff and Soil Loss

Retention of cropresidue on the soil layer can alsogives physical soil protection against water and soil loss. Crop residues cause a reduction of sediment load in surface runoff during rainfall (Bertol *et al.*, 2007). Combined effect of minimum or no-tillage with residue cover in protecting soil aggregates and decreasing soil loss and surface runoff (Govaerts *et al.*, 2007a; Panachuk i *et al.*, 2011).

3. Soil Moisture Content

One of the main benefits of surface residue cover ismoisture retention, in terms of yield increase, in rainfed climates where crop production is limited by soil moisture. Residue retention has been demonstrated to slow down runoff and reduce evaporation, conserving greater soil water content and resilience in drought-prone areas (Verhulst *et al.*, 2011b). Since residues also reduces surface compaction and sealing, infiltration rates for zero-tillage under full irrigation can be significantly higher than under conventional tillage (Verhulst *et al.*, 2011a). Effects of retention of crop residue on soil water some negative consequences also reported in certain environments. In semi-arid areas with small and frequent rainfall, residue cover can intercept rainfall and increase subsequent evaporation (Kozak *et al.*, 2007). Conversely, in areas with excess rainfall, residue retention can lead to excessive soil moisture and water logging (Rusinamhodzi *et al.*, 2011).

4. Soil Temperature

Retaining residues on the soil surface has been noted to decrease daytime soil temperature (Verhulst *et al.*, 2010a). In hot condition of tropical climates, this effect is beneficial because soil temperature may be too high for optimum plant growth whereas in cooler climates the effect can be detrimental to plant development. The effect of crop residue on soil temperature is controlled by the color of the residue. Dark residues resulted in higher mid-day temperatures compared to lighter colored residues (Sharratt and Campbell, 1994).

Effect of Crop Residue on Soil Chemical Properties

1. Soil Organic Carbon

Crop residue gives directly to soil organic matter and its decomposition is the initial stage in the humus formation process leading to C storage. Incorporation of crop residue can improve soil quality by increasing soil organic matter (Chen *et al.*, 2006). Soil organic carbon is important indicator of soil quality and agricultural sustainability because it improves soil aggregate stability and soil water retention, and provides a reservoir of soil nutrients (Alijani *et al.*, 2012) reported thatabsence of crop residue from agricultural lands, particularly associated with conventional tillage increases loss of soil organic carbon.

2. Soil pH

Retention of crop residues affect soil pH due to change in soil pH is related to the chemical composition of the residue and soil properties (Butterly *et al.*, 2011). Legume residues contributes greater effect on pH compared to residue with lower content such as wheat due to high in ash alkalinity and nitrogen in legume residues (Xu and Coventry, 2003).Changes in soil acidity from crop residue are combination effect with the concentration of organic anions in the residue and the nitrogen content of the residue (Butterly *et al.*, 2011).

3. Nutrient Availability

Crop residues can influence the availability of nitrogen to the crop. Addition of legume residues with a low C/N composition can result in nitrogen mineralization, whereas cereal residues with a high C/N composition can temporarily immobilization of nitrogen during the decomposition process (Govaerts *et al.*, 2006b).Retention of residue increases the concentration of phosphorus in the top soil. This can be attributed to redistribution of phosphorus mined from the deeper soil layers (Zibilske *et al.*, 2002).

4. Cation Exchange Capacity

Cation exchange capacity (CEC) is the capacity of the soil to hold cations for exchange with the soil solution and is an important indicator of soil fertility. Retention of soil residue increases soil organic matter content and thus increases the soil's pH dependent CEC. Lal (1997) observed that if residue retention increased CEC in both condition when residues were retained on the surface and incorporated, the increase in CEC was greatest when residues were retained on the soil surface.

Effect of Crop Residue on Soil Biological Properties

Effects of crop residue on soil microorganisms, the most abundant and diverse group of soil organisms (Roesch, 2007) and earthworms, the soil macrofauna integrally involved in soil aggregation processes. Both organisms are well support essential soil ecosystem functions, such as nutrient and soil carbon cycling, disease suppression, organic matter degradation, and soil structure (Wardle, 1995).

1. Microbial Activity

Soil microbial populations, including bacteria and fungi, are important actors in maintaining soil ecosystem functioning and sustainability (Paul, 2014). Soil microbial biomass (SMB) is defined as the living part of soil organic matter (Salinas-Garcia *et al.*, 2002). soil microbial biomass is considered an important indicator of soil quality. (Karlen *et al.*, 1997) because it is a source and reservoir ofbiologically available nutrients and promotes soil aggregation and structural formation (Angers *et al.*, 1992).The soil microbial biomass reflects the soil's ability to store and cycle nutrients (C, N, P and S) and organic matter (Carter *et al.*, 1999) and plays an important role in physical stabilization of aggregates (Franzluebbers *et al.*, 1999). Lou *et al.*, 2011, reported that in comparing treatments with straw retained and straw removed, they found significantlyhigher microbial biomass C levels when straw wasretained because of improved C and N contents, increased soilmoisture and porosity, and decreased soil temperature caused bythe residue cover. Incorporation of crop residues in the soil increases soil temperatureand aeration, creating favorable conditions for microorganismsand greater contact between them and the residues, whichlead to higher decomposition rates and overall SOC loss (Coppens *et al.*, 2007).

2. Earthworms

Burrowing and feeding activity of earthworms have numerous beneficial effects on overall soil quality for crop production. Earthworms are known as ecosystem engineers due to the species profound impacts on soil habitat: earthworms change soil chemical, physical, and structural properties, which subsequently impact soil biota and ecosystem functioning (Lemtiri *et al.*, 2015). They directly affect C and N cycles by consuming, storing, and cycling nutrients through their biomass (Whalen *et al.*, 1999). If crop residue is removed, earthworms lose their food source. So, use of a cover crop that is left in the field or removal of only part of the crop are ways to feed earthworms.Earthworms have been observed to respond positively to crop residue retention and minimum soil disturbance. The cooler soil temperature, improved soil structure, and food resource provided by crop residue retained on the surface can lead to increases in earthworm number and biomass (Chan and Heenan, 2006). So, Depending on the quality and quantity of residues incorporated in the soil versus that left on the surface, tillage may inhibit or enhance earthworm populations (Zaller *et al.*, 2004).

Conclusion

From the above point the influence of various crop residue treatments on soil properties, it could be concluded that crop residues recycled all the essential plant nutrients and ultimately increases the soil health or quality. Crop residueinfluenced the soil physical parameters favorablyvarious benefits of residue retention such as increased topsoil aggregation and protection from erosion, soil loss and surface compaction are increased when the residues are retained on the surface rather than incorporated. Therefore in regions with high risk of soil erosion and humid tropical areas with high decomposition rates it is particularly important to promote surface retention of residue.Negative effects of crop residue retention are less common;

however, they have been reported in some environments. The most common negative effect is N mobilization which can reduce N availability to the crop in N limiting environments. Crop residue increases Soil macro- and microorganisms, including earthworms and microbial communities, *i.e.* potentially key indicators of soil health.The study also showed that returning crop residue to the soil is beneficial; in ameliorating soil chemical conditions. So, effect of crop residue on soil quality and health due to increases physical, chemical and biological properties of agricultural soils. This ultimately leads to a more efficient, sustainable, and economically viable system.

REFERENCES

Alijani, K.,Bahrani, M.J. andKazemeini, S.A. 2012. Short-term responses of soil and wheat yield to tillage, corn residue management and nitrogen fertilization. *Soil Till Res*. 124: 78-82.

Angers, D.A., Peasant, A.andVigneux, J. 1992. Early cropping induced changes in soil aggregation, organic matter, and microbial biomass. *Soil Sci. Soc. Am. J.* 56, 115–119.

Bertol, O.J., Rizzi, N.E., Bertol, I. and Roloff, G. 2007. Soil and water loss and quality of surface runoff associated with interrill erosion in no-tillage area treated with chemical and organic fertilizers. *Rev. Bras. Cienc. Solo* 31, 781–792.

Butterly, C.R., Kaudal, B.B., Baldock, J.A. and Tang, C. 2011. Contribution of soluble and insoluble fractions of agricultural residues to short-term pH changes. *Eur. J. Soil Sci.* 62, 718–727.

Carter, M.R. 1992. Influence of Reduced Tillage Systems on Organic-Matter, Microbial Biomass, Macro-Aggregate Distribution and Structural Stability of the Surface Soil in a Humid Climate. *Soil Till. Res*. 23, 361-372.

Chan, K.Y. and Heenan, D.P. 2006. Earthworm population dynamics under conservation tillage systems in south-eastern Australia. *Aust. J. Soil Res.* 44, 425–431.

Chen, J., Yu, Z., Ouyang, J. and van, Mensvoort MEF 2006. Factors affecting soil quality changes in the North China Plain: a case study of Quzhou County. *AgricSyst* 91: 171–188.

Choudhury, S.G., Srivastava, S., Singh, R., Chaudhari, S.K., Sharma, D.K., Singh, S.K. and Sarkar, D. 2014. Tillage and residue management effects on soil aggregation: organic carbon dynamics and yield attribute in rice–wheat cropping systemunder reclaimed sodic soil. *Soil Tillage Res.* 136, 76–83.

Coppens, F., Garnier, P., Findeling, A., Merckx, R. and Recous, S. 2007. Decomposition of mulched versus incorporated crop residues: modelling with PASTIS clarifies interactions between residue quality and location. *Soil Biol. Biochem.* 39, 2339–2350.

FAO, 2011. Soil Health: Technologies that Save and Grow. FAO, Rome.

Franzluebbers, A.J., Haney, R.L., Hons, F.M. and Zuberer, D.A. 1999. Assessing biological soil quality with chloroform fumigationincubation: Why subtract a control? *Can. J. Soil Sci.* 79, 521-528.

Fuentes, M., Govaerts, B., De León, F., Hidalgo, C., Dendooven, L., Sayre, K.D. andEtchevers, J. 2009. Fourteen years of applying zero and conventional tillage, crop rotation and residue management systems and its effect on physical and chemical soil quality. *Eur. J. Agron.* 30, 228–237.

Govaerts, B., Sayre, K.D., Ceballos-Ramirez, J.M., Luna-Guido, M.L., Limon-Ortega, A., Deckers, L. and Dendooven, L. 2006b. Conventionally tilled and permanent raised beds with different crop residue management: effects on soil C and N dynamics. *Plant Soil.* 280, 143–155.

Govaerts, B., Sayre, K.D., Lichter, K., Dendooven, L. and Deckers, J. 2007a. Influence of permanent raised bed planting and residue management on physical and chemical soil quality in rain fed maize/wheat systems. *Plant Soil.* 291, 39–54.

Hargrove, W.L. 1991. Crop residue management in the Southeast. Crop Residue Management forConservation, Lexington, KY, *Soil and Water Conservation Society.*

Hiel, M.P., Chélin, M., Parvin, N., Barbieux, S., Degrune, F., Lemtiri, A., Colinet, G., Degré, A., Bodson, B. andGarré, S. 2016. Crop residue management in arable cropping systems under a temperate climate. Part 2: Soil physical properties and crop production. A review. Biotechnol. *Agron. Soc. Environ.* 20(S1): 245-256.

Karlen, D.L., Mausbach, J.W., Doran, J.W., Cline, R.G., Harris, R.F. and Schuman, G.E. 1997. Soil quality: a concept, definition, and framework for evaluation. *Soil Sci. Soc. Am. J.* 61, 4–10.

Kozak, J.A., Ahuja, L.R., Green, T.R. and Ma, L., 2007. Modelling crop canopy and residue rainfall interception effects on soil hydrological components for semi-arid agriculture. *Hydrol. Process.* 21, 229–241.

Lal, R., 1997. Long-term tillage and maize monoculture effects on a tropical Alfisol in western Nigeria. II: Soil chemical properties. *Soil Till. Res.* 42, 161–174.

Larson, W.E. and Pierce, F.J., 1994. The dynamics of soil quality as a measure of sustainable management. In: Doran, J., Coleman, D., Bezdicek, D., Stewart, A. (Eds.). *Soil Science Society of America, Inc., Special Publication. Number 35, Madison, Wisconsin,* pp. 37–51.

Lemtiri, A., Degrune, F., Barbieux, S., Hiel, M.P., Chélin, M., Parvin, N., Vandenbol, M., Francis, F. andColinet, G. 2016. Crop residue management in arable cropping systems under temperate climate. Part 1: Soil biological and chemical (phosphorus and nitrogen) properties. A review. *Biotechnol. Agron. Soc. Environ.* 20(S1): 236-244.

MNRE (Ministry of New and Renewable Energy Resources) 2009.Govt.of India, New Delhi. *www.mnre.gov.in/biomassrsources.*

Panachuki, E., Bertol, I., Alves Sobrinho, T., Sanches de Oliveira, P.T. andBicca Rodrigues, D.B.B., 2011. Soil and water loss and water infiltration in red latosol under different management systems. *Rev. Bras. Cienc. Solo* 35, 1777–1785.

Paul, E.A., ed., 2014. Soil microbiology, ecology and biochemistry. London: Academic Press.

Roesch,L.F.W. 2007. Pyrosequencing enumerates and contrasts soil microbial diversity. *roes* 1(4), 283290.

Rusinamhodzi, L., Corbeels, M., van Wijk, M.T., Rufino, M.C., Nyamangara, J.andGiller, K.E. 2011. A meta-analysis of long-term effects of conservation agriculture on maize grain yield under rainfed conditions. *Agron. Sustain. Dev.* 31, 657–673.

Salinas-Garcia, J.R., Velazquez-Garcia, J.D.J., Gallardo-Valdez, M., Diaz-Mederos, P., Caballero-Hernandez, F., Tapia-Vargas, L.M. and Rosales-Robles, E. 2002. Tillage effects on microbial biomass and nutrient distribution in soils under rain-fed corn production in central-western Mexico. *Soil Till. Res.* 66, 143–152.

Sharratt, B.S.and Campbell, G.S. 1994. Radiation balance of a soil-straw surface modified by straw color. *Agron. J.* 86, 200–203.

Singh, Y. and Sindhu, H.S., 2014. Management of cereal crop reidues for Sustainable rice-wheat production system in the indo-gangatic plains of india. *Proc Indian NatnSciAcad***80** (1) pp. 95-114.

Soon, Y.K. andLupwayi, N.Z. 2012. Straw management in a cold semi-arid region: impact on soil quality and crop productivity. *Field Crops Res.* 139, 39–46.

Turmel, M.S., Speratti, A., Baudron, F.,Verhulst, N. and Govaerts, B.2015.Crop residue management and soil health: A systems analysis.*Agricultural Systems* 134, 6–16.

Verhulst, N., Carrillo-Garcia, A., Moeller, C., Trethowan, R., Sayre, K.D. and Govaerts, B. 2011a. Conservation agriculture for wheat-based cropping systems under gravity irrigation: Increasing resilience through improved soil quality. *Plant Soil* 340, 467–479.

Verhulst, N., Nelissen, V., Jespers, N., Haven, H., Sayre, K.D., Raes, D., Deckers, J. and Govaerts, B. 2011b. Soil water content, maize yield and its stability as affected by tillage and crop residue management in rainfed semi-arid highlands. *Plant Soil* 124, 347–356.

Wardle, D.A. 1995. Impacts of disturbance on detritus food webs in agro-ecosystems of contrasting tillage and weed management practices. *Adv. Ecol. Res.*, 26, 105-185.

Whalen, J.K., Parmelee, R., McCartney, D. and Vanarsdale, J. 1999. Movement of N from decomposing earthworm tissue to soil, microbial and plant N pools. *Soil Biol. Biochem.* 31, 487–492.

Wilhelm, W.W., Doran, J.W.and Power, J.F. 1986. Corn and soybean yield response to crop residue management under no-tillage production systems. *Agron. J.* 78, 184–189.

Wilhelm, W.W., Johnson, J.M.F., Karlen, D.L. and Lightle, D.T. 2007. Corn stover to sustain soil organic carbon further constrains biomass supply. *Agron. J.* 99, 1665–1667.

Xu, R.K. and Coventry, D.R. 2003. Soil pH changes associated with lupin and wheat plant materials incorporated in a red-brown earth soil. *Plant Soil* 250, 113–119.

Zaller, J.G. andKöpke, U. 2004. Effects of traditional and biodynamic farmyard manure amendment on yields, soil chemical, biochemical and biological properties in a long-term field experiment. *Biol. Fertil. Soils,* 40(4), 222229.

Zibilske, L.M., Bradford, J.M. and Smart, J.R. 2002. Conservation tillage induced changes in organic carbon, total nitrogen and available phosphorus in a semi-arid alkaline subtropical soil. *Soil Till. Res.* 66, 153–163.

Conservation Agriculture Technologies *Pages 77–88*
Editor: **Rahul Kumar, Abhinandan Singh and Pankaj Kumar Ojha**
Published by: **BIOTECH BOOKS, NEW DELHI**

7 Prospects of Organic Farming in Horticulture

Arunima Tripathi[1] and Lokesh Bora[2]

[1]Assistant Professor, R.M.D. College of Agriculture and Research Station, Indira Gandhi Krishi Vishwavidyalaya – 492 012
[2]Department of Fruit Crops, Horticultural College and Research Institute Tamil Nadu Agricultural University, Coimbatore – 641 003, Tamil Nadu

Organic horticulture is the science and art of growing fruits, vegetables, flowers, or ornamental plants by following the essential principles of organic agriculture in soil building and conservation, pest management, and heirloom variety preservation. Instead of the plough, horticulture makes use of human labour and gardener's hand tools, although small machine tools like rotary tillers are common now. Organic farming provides a real opportunity for horticultural crop production, expend the food market worldwide (organically produced food has great demand in foreign countries there is major deficiency of home produced organic fruit and vegetable in certain countries like Island), has lower cost of production and provide good price premium which may cover reductions in yield during early phase of organic farming system.

Introduction

Organic farming is very much native to Indian land. One has to refer India and China while writing about organic farming. As farmer of these countries are sustaining themself based on their traditional farming based on organic practices since long time. The modern concept of organic farming combines the tradition, innovation and science. Although, history states that the movement for organic way of life recognized in 1905, it could gain ground after realizing the ill effects of modern agriculture in the late 1990's. Organic fruit production is a system devoid of the use of any chemical or genetically modified inputs, where the biological potential of the soil, organic sources and undergroundwater resources are conserved and protected by adopting suitable cropping pattern. Organic farming is very much native to this land. Many practices such as bio-dynamic farming, homa farming, panchagavya krishi, Rishi krishi and Natueco farming are associated with organic agriculture. With increasing the awareness regarding health the demand of qualitative and healthy agricultural products increased rapidly that is why different new method of agricultural farming has been emerged.The main aim of organic fruit production are includinglong-term fertility of soils, avoidance ofall forms of pollution that may result from chemical based farming, production of fruits of high nutritional quality and sufficient quantity and to reduce the use of energy in agricultural practice to a minimum level. Most of the organic fruit growers have expressed that lack of support price for organically grown crops and marketing infrastructure as the major constraint in promotion of organic agriculture.

Advantages of Organic Farming in India's Rural Economy

1. Organic fruits generally fetching 20 - 30 per cent higher price than non-organic fruits.
2. Capital investment is less than the chemical farming.
3. Minimize energy consumption by 30 per cent per unit of land.
4. Less dependence on money lenders (small farmers worldwide commit suicide due to increasing debt).
5. Traditional knowledge. (Chandrashekar, 2010)

Impact of Chemical Based Practices

1. Compaction of soil structure because of use of heavy machinery in field.
2. Low organic matter content.
3. Poor water holding capacity.
4. Increase in salinity and sodicity.
5. Adverse effect of chemicals on flora and fauna of soil.
6. Residual toxicity of chemical fertilizers and pesticides.

Principles of Organic Farming

1. The Principle of Health

Organic farming should sustain and enhance the health of the ecosystem for soil, plant, animal and human beings.

2. The Principle of Ecology

Organic farming should be based on living ecological processes and recycling. Nourishment and well-being are achieved through the ecology of the specific production environment.

3. The Principle of Fairness

Organic farming should build relationships that ensure fairness with regard to the common environment and life opportunities. Farming should ensure fairness at all levels and to all parties – farmers, workers, processors, distributors, traders and consumers.

4. The Principle of Care

Organic farming should be managed in a responsible and precautionary manner to protect the health and well-being of current and future generations and the environment. It states that precaution and responsibility are the key concerns in management, development and technology choices in organic agriculture.

How to Control Quality of Organic Produce

There are four keys to control quality of the produce, namely Accreditation, Standards, Inspection and Certification.

1. Accreditation

It guarantees that the certification program is competent to carry out specific tasks. Authoritative body defines policies, standards and checks whether a certification system is operating according to standards.

2. Standards

It defines production methods, not the product quality. It is minimum requirements, not "best practice". Standards are continuously developed and dynamic in nature. It can be International, National or regional standards.

3. Inspection

On-site visit to verify the performance of an operation is in accordance with specific standards.

4. Certification

Certification is a written confirmation that a process or product is in compliance with prescribed standards. Certification granted on prescribed form for a period of one year.

Global Organic Production Scenario

1. The global trade during 2013-14 was USD 60 billion (Rs. 3, 60,000crores) and may touch USD 100 billion (Rs. 6, 00,000 crores) within the next five years.
2. Annual market growth of global organic food has been expanded day by day and in 2010 the annual growth was noticed 23 per cent.
3. Organic farming is currently practiced in 162 countries covering 37.5 million ha (2012, IFOAM).
4. Australia has highest organic agriculture land (12.2 million hectare) followed by Argentina (2.8 million hectare).

Distribution of Organic Agriculture Land by Region (2012)

The regions with the largest areas of organically managed agricultural land are Oceania (12.1 million hectares of 33 per cent of the global organic farmland), Europe (10.6 million hectares or 29 per cent of the global organic farmland) and

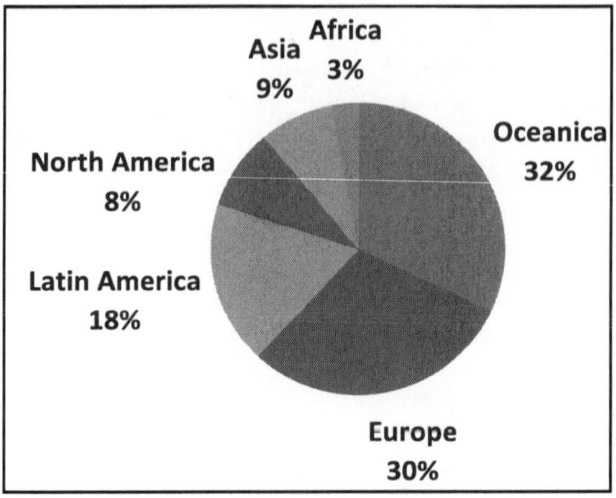

Figure 7.1: Distribution of Organic Agriculture Land by Region (2012) (IFOAM Survey, 2014).

Latin America (6.8 million hectares or 23 per cent). On a global level, the organic agricultural land area increased by three per cent compared with 2010.

Organic Production and Export in India

India exported 135 products last year (2013-14) with the total volume of 194088 MT including 16322 MT organic textiles. The certified area includes 15 per cent cultivable area with 0.72 m haand rest 85 per cent (3.99 mha) is forest and wild area for collection of minor forest produces. The total area under organic certification is 4.72 million Hectare (2013-14). Exported items are soybean (70 per cent), cereals and millets (6 per cent), processed food products (5 per cent), Basmati rice (4 per cent), sugar (3 per cent), tea (2 per cent), pulses and lentils (1 per cent), dry fruits (1 per cent) and spices (1 per cent).

Increase in export with more farmers shifting to organic farming. Organic food products in India are priced about 20 - 30 per cent higher than non-organic food products. Though the salaries in India are increasing rapidly, the domestic market is not sufficient to consume the entire organic food produced in the country. As a result, exports of organic food are the prime aim of organic farmers as well as the government.

India produced around 1.24 million MT of certified organic products which includes all varieties of food products namely Sugarcane, Cotton, Oil Seeds, Basmati rice, Pulses, Spices, Tea, Fruits, Dry fruits, Vegetables, Coffee and their value added products. The production is not limited to the edible sector but also produces organic cotton fiber, functional food products *etc.*

States like; Uttarakhand, Karnataka, Madhya Pradesh, Maharashtra, Gujarat, Rajasthan, Tamil Nadu, Kerala, Nagaland, Mizoram and Sikkim have been promoting organic farming.

Among all the states, Madhya Pradesh has covered largest area under organic certification followed by Himachal Pradesh and Rajasthan.

1. India is best known as an exporter of organic tea and also has great export potential for many other products. Other organic products for which India has a niche market are spices and fruits.
2. Major organic fruit crop produced in India are mango, banana, pineapple, passion fruit, orange, cashew nut and walnut.Among the fruit crops bananas, mangos and oranges are the most preferred organic products.
3. Many states of India have promoted policies and programmes on organic farming. The states of Uttarakhand and Sikkim have declared their states as 'organic states'.
4. Karnataka and Andhra Pradesh formulated organic policies and Maharashtra, Tamil Nadu, Kerala and West Bengal supported public: private partnerships for promoting organic farming.In Gujarat, organic production of sapota, banana and coconut was found to be more profitable. Arunachal Pradesh can be explored as a destination for organic citrus fruit production.
5. The North-Eastern region (eight states) cultivating large number of fruit crops like pineapple, banana, citrus, litchi, mango, passion fruit, kiwifruit, *etc.* are grown in this region.
6. Except for a few fruit crops like mandarin, pineapple, passion fruit and litchi, other crops are grown in the homestead – about 60 per cent of the horticultural production has been contributed by homestead gardens and more than 90 per cent of the gardens are organically based.
7. 'The National Commission on Farmers'has identified organic farming as a tool for second green revolution in the rain-fed and hilly areas.
8. In most of the importing countries, organic shares in fresh fruit sales are estimated at about 5 to 7 per cent. There is a growing demand for organic fruits in United States, EU, Japan and Australia and the main suppliers will be producers in developing countries.
9. Currently, India exports mainly mango, grapes, apple, citrus and some other fruits like banana, sapota and litchi to Middle East, Europe (particularly UK) and to some extent the Far East (Singapore, Malaysia, *etc.*). While the key organic fruit markets are US, EU, and Japan.
10. EU member states, especially UK, The Netherlands and Germany have a good demand for organic mangoes, which can be exploited by India. Processed organic mango in the form of pulp, juice, puree, chutneys, jams and pickles also has a similar export destination profile.

Indian organic banana exports (dry banana and banana puree) are negligible as compared to world trade. Need strategies- firstly, target the processed organic banana market (banana pulp, purees and concentrates) and secondly focus on the geographically closer Japanese market and the EU.India is the 4th largest producer of pineapple accounting for about 11 per cent of total world output. <1 per cent of the produce is exported.

Grapes are India's largest exported fruit; the main target destination market for Indian organic grapes is the EU, specially the UK and the Netherlands.Passion fruit is popular in western countries (lemony tart flavour), mainly consumed as a fruit juice or as an ingredient in multi-vitamin drinks. It is grown organically in North Eastern States. There is negligible export because of lack of world-class processing and packaging facilities.

India is the second largest producer of litchi, Fruit harvesting in Australia, Mauritius, South Africa and Madagascar falls between November and March, The UK, Germany and France which normally import litchi from Madagascar and South Africa, can be very well supplied with Indian organic litchi during the summer months.

Regulatory Bodies for Promotion and Certification of Organic Produce

For quality assurance the country has internationally acclaimed certification processin place for export, import and domestic markets. National Programme on Organic Production (NPOP) defines the regulatory mechanism and is regulated under two different acts for export and domestic markets. National Program on Organic Production (NPOP) was launched in the year 2000, which was formally notified in October, 2001. The document provides information on standards for organic production, systems criteria, and procedures for accreditation of Inspection and Certificationbodies, the national organic logo and the regulations governing its use.

National Accreditation Body (NAB). 20 accredited certification agencies in India. The NAB is responsible for giving approval on the authorization of an applicant Inspection and Certification Agency. On approval given by NAB, the APEDA issues a Certificate of Authorization.In India, (APEDA) is the controlling body for organic certification for export. The standards and procedures have been formulated in harmony with international standards such as those of Codex and IFOAM.Indian organic certifying bodies have received European Commission (EC) 'equivalence' status exempting Indian organic products from further certification by European agencies, which is likely to bring down the cost of organic exports.

Gap for Organic Fruit Production

Presently, GAP is formally recognized in the international regulatory framework for reducing risksassociated with the use of pesticides, taking into account public and occupational health, environmental and safety considerations. Some of the major risk and minimizing measures are highlighted below:

Pre-Planting Measures

Site selection

It should be selected on the basis of land history, previous manure applications and crop rotation. Soil should be free from heavy metals and other toxic chemicals.

S. No.	Centre	States under Jurisdiction of different centres for training
1.	National Centre of Organic Farming, Ghaziabad	Delhi, Uttar Pradesh, Uttrakhand and Rajasthan. All India for referral QC samples and All India level programmes.
2.	Regional Centre of Organic Farming, Bangalore	Karnataka, Kerala, Tamilnadu, Puduchheri, Lakshdweep
3.	Regional Centre of Organic Farming, Bhubaneshwar	Orissa, W. Bengal, Bihar, A & N islands and Sikkim
4.	Regional Centre of Organic Farming, Imphal	Arunachal Pradesh, Assam, Manipur, Meghalaya, Mizoram, Nagaland, Tripura
5.	Regional Centre of Organic Farming, Hisar	Haryana, Punjab, Himachal Pradsh and J & K
6.	Regional Centre of Organic Farming, Jabalpur	Madhya Pradesh, Chattisgarh, Jharkhand
7.	Regional Centre of Organic Farming, Nagpur	Maharashtra, Andhra Pradesh, Gujarat, Goa, UT of Daman, Diu and & N Haveli

Working Jurisdiction of NCOF and RCOFs (www.dacnet.nic.in/ncof).

Manure Handling and Field Application

Livestock manure can be a valuable source of nutrients. Proper composting of manure and its incorporation into soil prior to planting is recommended rather than top-dressing of plants. Manure should be applied at the end of the season or two weeks before planting.

Selection of Appropriate Crop

Selection of material for planting is determined by a number of factors. These include: growing conditions, pest and diseases resistance, fruit quality parameters (such as size, firmness and keeping ability), maturity dates, market requirements *etc.*

Production Measures

Irrigation Water Quality

Ideally, water used for irrigation should be free from pathogens and heavy metals. However, potable water or municipal water is not feasible for extensive use for crop production. Surface water used for irrigation should be quarterly tested in laboratory for pathogen.

Irrigation Methods

Drip irrigation method should be used, whenever possible to reduce the risk of crop contamination because the edible parts of most crops are not wetted directly. Plant disease levels also may be reduced and water use efficiency is maximized with this method.

Plant Protection

Pests are controlled using various methods *i.e.*, cultural, physical and mechanical means. Chemical pesticides are prohibited. A number of bio- pesticides are available which can be used judicially. So, an integrated pest and disease management (IPDM) approach for pest and disease management is followed for better result. All applications should be documented. The minimum interval between such treatments and harvest should be consistent with the labeling.

Field Sanitation and Animal Exclusion

Farmers should stay out of wet fields to reduce the spread of plant or human pathogens.

Tractors that were used in manure handling should be cleaned prior to entering produce fields.

Animals, including poultry or pets should not be allowed to roam in crop areas, especially close to harvest time.

Worker Facilities and Hygiene

Ideally, farm workers should be provided clean, well-maintained and hygienic toilet facilities around the farming areas. Farmers should get proper training to make them understand the relationship between food safety and personal hygiene.

Harvest

Clean Harvest Aids

All crop containers have to washed and rinsed under high pressure. All crop containers should be sanitized before harvest.

Worker Hygiene and Training

Good personal hygiene is particularly important during the harvest of crops. Sick employee's orthose with contaminated hands can spread pathogens to produce.

Post-Harvest Handling

Worker Hygiene

Hands can contaminate fresh fruits with harmful microbes and other chemicals. Packing area shouldbe cleaned and sanitized. Supply liquid soap in dispensers, potable water, and single-use papertowels for hand washing. Workers should be properly educated about the importance of restroomuse and proper hand washing.

Sanitize Packinghouse and Packing Operations

Loading, staging, and all food contact surfaces should be cleaned and sanitized at the end of each day. Wash, rinse and sanitize the packing line belts, conveyors, and food contact surfaces at the end of each day to avoid buildup of harmful microorganisms. Packaging material should be free from chemical and there should be labeled with all information with an organic logo.

Transportation of Produce from Farm to Market

Proper cleanliness of the transportation vehicles should be ensured before loading. Make sure that fresh fruits and vegetables are not shipped in trucks which have carried live animals or harmful substances.

India-Strength

- Wide variety of fruits and vegetables,
- Very less consumption of chemicals in India. These areas can be turned into organic
- Organic pockets existing in different parts of the country
- Farmers as well as Government interests are rising in organic
- Various niches in fruits and vegetables can be created
- Indian corporate investing in agribusiness especially in organic farming.

(*Source*: http: / /www.organicmonitor.com)

Major Challenges in Moving to Organic Production

- Approved Materials (Restriction on use of products -synthetic pesticides and fertilizers, Decertification).
- Certification (Product claiming "organic" be certified by a third party body, Costly, requires careful and extensive record keeping, growers administrative workload increases).
- Labor (Tree fruit production- labour-intensive, need more labour 15 per cent more charge).
- Storage and packing (proper care, no chemical treatment)
- Horticultural Aspects (weed control, pest and disease control and fertility management).

How Organic Fruit Production can be Promoted in India

From Policy perspective

- Policy in favour of organic production, encouraging production of bio-inputs, changes in cropping pattern in favour of organic practices

From Farmers Perspective

- Existing subsidy structure (Promotes chemical fertilizers and pesticides)
- Grant assistance (converting to organic, certification fees)
- Special insurance scheme
- Promote contract farming based on organic production.

From NGO's perspective

- Public private partnership - successful story of Spice Board's involvement of NGO's in organic spice production

From Traders Perspective

- Organic agricultural export market potentials and premium price in different parts of the world
- Establishment of Special Organic Agriculture Trade Zone (OATZ)
- Providing special tax relaxation for traders/exporters
- To enhance the domestic organic agricultural market government need to make efforts towards promotion of organic fruit processing industries (Financial and subsidy assistance).

From Green Input Producers Perspective

- Popularization of existing scheme for promotion of bio-inputs,
- Assistance in marketing of bio-inputs thorough govt. network
- Promotion of corporate research for organic agronomic practices, bio-control of diseases and pest, bio-fertilizers *etc.*

From Institutional Perspective

Requires inter-ministerial coordination among various ministries at the Centre for effectiveness of policies and programmes Such as:

- Ministry of Agriculture,
- Ministry of Commerce and Trade,
- Ministry of Environment,
- Ministry of Science and Technology,
- Ministry of Finance *etc.*

(*Source*: http: //www.fao.org/docrep/article/agrippa/658_en-07.htm)

Future Scope of Organic Fruit Production

- Organic fruit production has highly expanded and strengthens future prospects because of quality of organic fruits are far better than chemical fruit production without any environmental, soil losses.
- Organic farming continuously increasing by 25-30 per cent per year from last 10 years.
- Organic fruit production is an economical and reliable option of profitable livelihood having bright scope in future. (Yadav *et al.*, 2010)

Other Limitations/Weakness

- Short shelf life varieties.
- Lack of farmer awareness about agricultural practices, products and technologies for organic farming.
- Quality not competitive in the domestic and international market.
- Lack of market information and intelligence.

- ☆ Global marketing research lacking.
- ☆ Inadequate post-harvest management and related specialized infrastructure to support organic fruit production.
- ☆ R&D base in organic fruit production lacking.
- ☆ Unavailability of inputs used in system.
- ☆ Lack of proper infrastructure (roads, cold storage facilities affect the cost, quality and reach of producers).
- ☆ No access to international market.

Conclusion

- ☆ Organic fruit production can provide quality fruit without adversely affecting the soil's health and the environment.
- ☆ Organic farming helps in rejuvenating the degraded soil and ensure sustainability of crop production. So it should be promoted.
- ☆ Negligible amount of organic fruits are exported now more potential is there as more demand in market, and consumers are ready to pay premium prices.
- ☆ Further research work on organic fruit production can envisage cheap production cost for the farmers and pave the way to aimed success.

REFERENCES

Bhattacharya, P. and Chakraborty, G. 2005. Current status of organic farming in India and other countries.*Indian Journal of Fertilizers* **1**(9): 111-123.

Chandrashekar, H. M. 2010. Changing scenario of organic farming in India: An overview.*International NGO Journal*.**5 (1)**: 034-039.

Choubey, Manesh. 2006. Organic farming in Northeast India: A Case Study of Organic Orange Cultivation in Arunachal Pradesh. Agricultural Economics Research Review.19: 236-245.

http: //www.apeda.gov.in

http: //www.agritech.tnau.ac.in

http: //www.dacnet.nic.in/ncof

http: //www.fao.org/docrep/article/agrippa/658_en-07.htm

http: //www.icarneh.ernet.in

http: //www.ifoam.bio/

http: //www.ncof.dacnet.nic.in

http: //www.organicmonitor.com

http: //www.organicworld.net/yearbook-2014.html

IFOAM (2014), Workshop on Alternatives on Certification for Organic Production, April 2014.

Lammerts van Buerena, E.T., Jones, S.S., Tamm., L., Murphy., K.M. Myers, J.R., Leifert, C and M.M. Messmer.2010.The need to breed crop varieties suitable for organic farming, using wheat, tomato and broccoli as examples: A review. *Wageningen Journal of Life Sciences.***58:** 193– 205.

Mitra, S.K., 2013. Organic Tropical and Subtropical Fruit Production in India – Prospects and Challenges. *Proc.* 4thInternational Symposium on Tropical and Subtropical Fruits. *Acta Hort.***975**: 303-309.

Munda, G. C., Das Anup and Patel, D. P.2010. Organic Farming in Hill Ecosystems – Prospects

and Practices. ICAR Research Complex for NEH Region (Report).

Ramesh, P., Panwar, N.R., Singh, A.B., Ramana,S., Srivastva, Rahul and SubbaRao. A.2010. Status of organic farming in India. *Current Science.***98**: 1190-1194.

Robyn Neeson. 2008. Organic fruit production. *Prime fact*(July), pp. 1-8.

Singh, Vishal.,Singh, Gurupreet and Verma, Deepak 2013. Organic Foods: Benefits and Market Demand. *Pop. Kheti.* **1(4)**: 71-78.

Conservation Agriculture Technologies *Pages 89–93*
Editor: **Rahul Kumar, Abhinandan Singh and Pankaj Kumar Ojha**
Published by: **BIOTECH BOOKS, NEW DELHI**

8 Rainfed Agriculture in India: Challenge and Importance

Reena Kumari[1], Babloo Sharma[2] and Pratibha Kumari[3]

[1]Agricultural Engineering Department, N.M. College of Agriculture, Navsari Agricultural University, Navsari – 396 450, Gujarat
[2]Department of Soil and Water Conservation, Faculty of Agriculture, Bidhan Chandra Krishi Viswavidyalaya, Mohanpur, Nadia – 741252, West Bengal
[3]Department of Soil Science and Agricultural Chemistry, Institute of Agricultural Sciences, Banaras Hindu University, Varanasi – 221 005, Uttar Pradesh

Introduction

Rainfed areas in India are highly diverse, ranging from resource rich areas with good agricultural potential to resource-constrained areas with much more restricted potential. The best land for agricultural production has been reducing by the time, due to high agricultural expansion. The population density is growing up, as a consequence the demand for land resources such as food, fuel and shelter has been increasing. There is a need for exploitation of land which is less suitable for agriculture, or land in less favorable climates. India ranks first among the rainfed agricultural countries of the world in terms of both extent (86 M ha) and value of produce. Due to low opportunities and higher population of landless households and agricultural laborers as well as low land and labor productivity, poverty is concentrated in rainfed regions. Agriculture is practised on 12 per cent of the total land area, hosting around 42 per cent of the global population. Most of this area, around 80 per cent, is under rainfed agriculture, which plays a predominant role in global food supply and water demand for food.

Eight hundred million people are food- insecure, and 166 million pre-school children are malnourished in the developing world. Producing enough food, and generating adequate income in the developing world to better feed the poor and reduce the number of those suffering will be a great challenge. This challenge is likely to intensify, with a global population that is projected to increase to 7.8 billion people in 2025, putting even greater pressure on world food security, especially in developing countries where more than 80 per cent of the population increase is expected to occur. Per capita water availability in India shown in Figure 8.1. Irrigated agriculture has been an important contributor to the expansion of national and world food supplies since the 1960s, and is expected to play a major role in feeding the growing world.

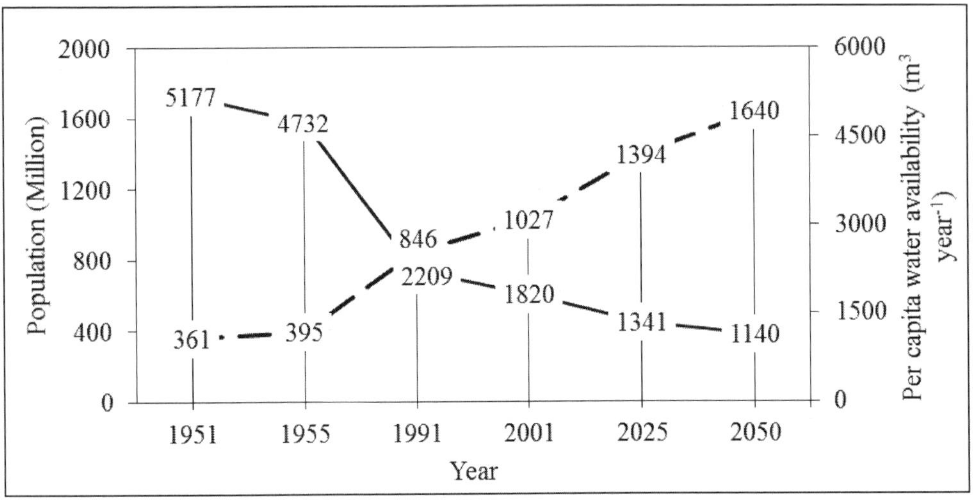

Figure 8.1: Per Capita Water Availability in India (*Source*: Government of India, 2009).

Out of concern over increasing water scarcity for irrigation, the role of water management and investments for irrigated agriculture and food security has received substantial attention in recent years. Rainfed areas currently account for 58 per cent of world food production. Given the importance of rainfed cereal production, insufficient attention has been paid to the potential of production growth in rainfed areas to play a significant role in meeting future food demand.

Challenges of Rainfed Agriculture

The most recent estimates put global rainfed croplands at 1.75 billion hectares at the end of the last millennium, which are about 5.5 times the irrigated areas of the world. The low and variable productivity of these lands is the major cause of poverty for 70 per cent of world's poor inhabiting these lands. On average, 1300 cubic meters per capita per year (m^3 cap^{-1} yr^{-1}) of freshwater are required to produce the food for a healthy diet of 3000 kcal cap^{-1} day^{-1} with a 20 per cent meat share. This means that >8000 km^3 yr^{-1} of water are consumed (*i.e.* evapotranspired on rainfed and irrigated land) to feed the current world population, and that an additional 5000

$km^3\ yr^{-1}$ will be required if the population rises to 10 billion in 2050 as suggested by the IPCC's SRES A2r scenario.

Rainfed agriculture plays an important role in the Indian economy. In India, 68 per cent net sown area (136.8 m ha) comes under rainfed agriculture. In India, some 48 per cent area is under food crops and 68 per cent area under nonfood crops. According to the National Rainfed Area Authority (NRAA) almost 50 per cent rural work force and 60 per cent livestock is dependent on rainfed agriculture. More than 177 districts of India are dominantly rainfed districts. Arguably, a significant increase in the form of groundwater irrigation is reported from some of these areas in the last few decades. Rainfed areas also match closely with the areas characterized by human poverty. Under various programs, the Government has invested almost thrice the amount in irrigation projects as compared to investments in rainfed areas. Subsidies continue to flow to farmers with irrigation facilities, in the form of drips, sprinklers or fertilizers. Under the National Food Security Programme, emphasis to crops like wheat and paddy is obvious. Various policy documents fall short on the manner of priority given to water, soil and especially to rainfed agriculture. This is the straightforward rationale to revitalizing rainfed agriculture in India.

The agricultural productivity has seen a rapid growth since the late 1950s due to new crop varieties, fertilizer use and expansion in irrigated agriculture. Food and crop demand is estimated to double in the next 50 years. According to a Comprehensive Assessment, it is possible to produce food – but it is probable that today's food production and environmental trends, if continued, will lead to crises in many parts of the world. The importance of rainfed sources of food weighs disproportionately on women, given that approximately 70 per cent of the world's poor are women. Agriculture plays a key role for economic development and poverty reduction, with evidence indicating that every 1 per cent increase in agricultural yields translates to a 0.6–1.2 per cent decrease in the percentage of absolute poor.

A major challenge for rainfed areas is that of protecting, conserving and intelligently using water resources especially considering a water management strategy in rainfed areas also holds the key to water management solutions in areas where largescale groundwater over-exploitation has occurred and both *kharif* and *rabi* cropping has become vulnerable to water shortages.

Climatic Characteristics of Rainfed Agriculture

The climate in these regions is characterized by complex climatic deficiencies, manifested as water scarcity for rainfed crop production. The climate is largely semi-arid and dry sub-humid with a short (occasionally intense) wet season followed by long dry season. Rainfall is highly unreliable, both in time and space, with strong risks of dry spells at critical growth stages even during good rainfall years. Inter-annual fluctuations are high due to monsoonal climate- characteristics of the atmospheric circulation and strong links to ENSO phenomenon in the Pacific Ocean. The rainfall in the tropics is seasonal with erratic distribution and generally comes as torrential downpours, resulting in large runoff and severe soil erosion. Groundwater levels are depleting in the region and most regions are facing water scarcity and drinking water shortages in the summer season. Soils of these areas are deficits in nutrient and the productivity in these areas is low. The water scarcity is largely due to improper management of water resources. Rainfall use efficiency in the rainfed areas is low (35 to 55 per cent) and through improved management options the productivity of rainfed agriculture could be doubled with enhanced water resources use efficiencies. Rainfed areas in India are highly diverse, ranging from resource rich areas with good agricultural potential to resource-constrained areas with much more restricted potential. Agriculture in rainfed areas continue to be constrained by inadequate water availability as monsoon rains is undependable both in time and amount. Barring this constraint, a substantial part of the rainfed lands have the greatest unused potential for growth. Rainfed agriculture is mainly and negatively influenced by intermittent dry spells during the cropping season. Delay in onset of monsoon season triggers shifts to other small duration crops tailored to the remaining length of the growing season rather than overall crop failure. Intensity of dry spells at other critical stages of crop growth, especially during the flowering/grain filling stage of the crops (even for 1 to 2 weeks), have large bearing on the potential yields at the farmer fields and several times leads to even crop failures. In order to achieve stability in yields, it is imperative to provide protective/critical/supplementary irrigation during these short critical periods.

Rainfed agriculture is practiced under a wide variety of soil type, agro-climatic and rainfall conditions ranging from 400 mm to 1600 mm per annum. It is estimated that 15 M ha of rainfed cropped area lies in arid regions and receives less than 500 mm rainfall, another 15 M ha is in 500-700 mm rainfall zone, and bulk of 42 M ha is in the 750-1100 mm rainfall zone. The remaining 20 M ha lies in 'above 1150 mm/annum' zone. As rainfed production is spread over different climatic regions, it offers great scope for raising a number of diversified crops. At the same time, potential of improving agricultural productivity under rainfed conditions thus also varies considerably. Earlier, the rainfed farming systems, because of its risky nature was dependent upon locally available inputs (seeds, manures, animal draft) and used to grow a number of crops, which were able to withstand drought-like situation. But over time, the cropping systems have changed and presently farmers in these rainfed areas have limited options and have stated cultivating high value crops which requires intensive use of costly inputs (chemical fertilizers/pesticides, hybrid seeds, life saving irrigation, farm energy *etc.*) and find it difficult to manage the resources on their own. Thorat (1993) use narrow rainfall thresholds with a relatively low maximum because they intended to focus on very dry (but not quite desert) areas and percentage of gross cropped area under irrigation (less than 25 per cent) and average annual rainfall (between 400 and 750 mm). The percentage of gross cropped area under irrigation (less than 10 per cent) and average annual rainfall between 375 and 750 mm.

Importance and Contribution of Rainfed Agriculture

Rainfed agriculture with nearly 58 per cent of the cultivated area contributes 40 per cent of the country's food production. Even after full irrigation potential of the country is realized, half of the cultivated area will continue to be under rainfed farming. Much of the acreage under coarse cereals (85 per cent), pulses (83 per cent) and oilseeds (70 per cent), substantial area under rice (42 per cent) and nearly 65 per cent of cotton area is rainfed (Table 8.1). Hence, it is necessary to increase the productivity of major rainfed crops to meet the ever increasing demand of food and fiber. Moreover, rainfed regions are home to about 40 per cent of the human and 60 per cent of the livestock population, and the performance of rainfed agriculture is critical to achieve and sustain higher growth in agriculture. Thus, rainfed agriculture assumes importance from the consideration of growth, equity and sustainability.

Table 8.1: Area Sown under Various Rainfed Crops and Percentage of Rainfed Area during 2008–09*

Crop	*Area Sown (M ha)*	*Rainfed Area (per cent)*
Rice	45.5	42
Coarse cereals	27.5	85
Jowar	7.5	91
Bajra	8.7	91
Maize	8.2	75
Pulses	22.1	83
Redgram	3.4	96
Bengal gram	7.9	67
Oilseeds	27.6	70
Groundnut	6.2	79
Rapeseed and mustard	6.3	27
Soybean	9.5	99
Sunflower	1.8	69
Cotton	9.4	65

* *Source*: Venkateswarlu and Prasad (2012).

REFERENCES

Govt. of India. 2009. Background note for consultation meeting with Policy makers on review of National Water Policy. Ministry of Water Resources. p. 50.

Thorat S K 1993. Technological Change and Regional Differentiation. New Delhi: Khama Publishers.

Venkateswarlu B. and Prasad J. V. N. S. (2012). Carrying capacity of Indian agriculture: issues related to rainfed agriculture. *Current Science,* 102(6): 882-888.

Conservation Agriculture Technologies *Pages* **95–110**
Editor: **Rahul Kumar, Abhinandan Singh and Pankaj Kumar Ojha**
Published by: **BIOTECH BOOKS, NEW DELHI**

9 Protection of Crops from Plant Pathogens in Sustainable Agriculture

Ankit Kumar Ghorai and Subrata Dutta

Department of Plant Pathology,
Bidhan Chandra Krishi Viswavidyalaya, Mohanpur – 741 252, West Bengal

The current 6.7 billion world population is expected to shoot up to 10 billion people by 2050. This 45 per cent increase in current world population will impose greater demand of increased food supply. Many agricultural lands are being transformed to industrial and human dwelling places. The target of Indian agriculture is to produce 345 million ton (mt) in 2030, which requires an addition of 5-6 mt annually. In addition to this the per capita availability of agricultural land is shrinking from 0.48 hectare in 1951 to 0.13 hectare in 2011 and will further decline to 0.08 hectare in 2035. Although pesticides have the potential to boost up production of food and fiber, but their use have proved to be hazardous to man and his environment. In this context, sustaining and enhancing the soil fertility, minimising the selection pressure over plant pathogens imposed by fungicides and improving the soil biological ecosystem will help us to bestow a healthy, productive and diversified ecosystem for agriculture to the coming generations.

Various methods for protection of crop diseases include the following in a sustainable eco-friendly way includes:

1. Growing Resistant Varieties

This is the most economic and "pain less" method protecting crops from plant pathogens in a sustainable way. Use of resistant varieties reduces the burden of spraying harmful fungicides and pesticides which in turn reduces the risk of residual effect in plants, soil and food chain. Further, in a developing country like

India it relieves the marginal farmers from bearing the cost of spraying and dusting crops. Extensive use of pesticides also reduces the population of beneficial micro-organisms in phylloplane and rhizosphere and leads to evolution of new virulent races of pathogen. Recent studies have found that resistant varieties harbor beneficial micro-organisms in their phylloplane and rhizosphere which in addition giving a combat to target pathogen, also helps in better growth of the crops.

1.1 Types of Host Resistance

Three types of resistance imparted by host includes non host resistance, true resistance and apparent resistance.

1.1.1 Nonhost Resistance

Non host resistance is a complete resistance because the nonhost plants fall completely outside the host range of certain pathogens. For example: Mango is completely resistant to *Ralstonia solanacearum* wilt.

1.1.2 True Resistance/True Resistance

This resistance is due to resistance genes inherited by plants.

1.1.3 Apparent Resistance

This is due to disease escape or tolerance.

1.1.3.1 Disease Escape

Crops may escape disease under following conditions:

(i) *Pythium* sp. causing damping off and pathogens causing powdery mildew are unable to attack crop plants in adult stages when the secondary growth has already occurred. While pathogens like *Botyrtis* sp., *Penicillium* sp. and *Alternaria* sp. cannot attack tender shoots and leaves of plant, as their incidence is observed during matured stages of crops.

(ii) Seeds and seedlings having high vigour and germination rate are avoided by many pathogens.

(iii) Low soil moisture keeps damping off incited by *Pythium* sp., *Phytophthora* sp. and *Plasmodiophora brassicae* (Club rot of crucifers) as these fungi requires adequate soil moisture for the spread of zoospores and oospore germination. While *Streptomyces scabies* (Common scab of potato) cannot attack potato under irrigated or wet soil conditions.

(iv) Plants escape diseases incited by *Pythium* and *Phytophthora* under high temperature conditions. For instance, *Pythium* requires 10°C for germination of oospores.

1.1.3.2 Disease Tolerance

Those varieties which despite of being infected by plant pathogen shows no loss in yield are said to show tolerance. For example Hisar Shyamal (H8) variety of brinjal is tolerant to little leaf, Sadabahar of chickpea is tolerant to wilt, Jawahar variety of pigeon pea is tolerant to *Phytophthora* blight and wilt, barley variety Proctor is tolerant to powdery mildew.

1.1.3.3 Durable Resistance

Growing of varieties having durable resistance helps to combat resistance breakdown. For example: mlo gene in barley against *Erysiphe graminis* f.sp. *hordei,* Sr26 against black stem rust of wheat (*Puccinia graminis tritici*), Lr34 against brown rust of wheat (*Puccinia recondita*) still are resistant after their introduction.

1.2 Methods of Breeding for Disease Resistance

Resistance to target pathogens can be harnessed by the following ways.

Table 9.1: Different Methods to Develop Plant Disease Resistant Varieties

Sl.No.	*Methods*	*Description*	*Examples*
1	**Selection**	Selection and introduction and multiplication of varieties that escape infection under field conditions when all the environmental conditions are conducive.	1. Kufri red potato is a slection from Darjeeling red round is resistant to late blight. 2. Cotton variety MCU-1 selected from Co4 is resistant to black arm.
2	**Introduction**	In case known varieties are not found to show resistance then proper screening of germplasms are done. Example: Resistance to wilt of watermelon, capsicum, linseed and blast of rice.	1. Ridley introduced from Australia, Kalyansona and Sonalika are secondary selections from CIMMYT introduced cultivars were resistant to rust of wheat. 2. Downey mildew resistant male sterile lines of Pearl millets (Tift 23A cytoplasm) were developed from African introductions.
3	**Mutations**	Spontaneous and induced mutations have been well utilized to fight back plant pathogens. Example:	1. Pusa 408 and Pusa 413 of chickpea resistant to Ascochyta wilt. 2. Pant Mung-2 resistant to YVMV and powdery mildew.
4	**Hybridization**	The most common method Backross- Transfer of disease resistance gene to an agronomically suitable variety.	
		Pedigree Method	Kalyasona, Sonalika,Malviya 206
		Somatic hybridistaion	*Solanum brevidens* + potato= Resistance to PLRV *Brassica nigra+ brassica napus* =Resistance to *Phoma lingum* and *Plasmodiophora brassicae*
5	**Somaclonal variations**	Variation observed in plant tissue culture has been utilized to grow various resistant varieties. Example: Sugarcane variety Ono resistant to Fiji disease, Smut and Sugarcane mosaic virus, tobacco varieties released against TMV, PVY and *Alternaria alternata*.	Ono of Sugarcane resistant to Fiji disease, smut and Sugarcane mosaic virus. Potato varities resistant to early and late blight. Tobacco varieties resistant to TMV and PVY.

Sl.No.	*Methods*	*Description*	*Examples*
6	**Genetic Engineering**	Transgenic plants expresses genes which by various mechanisms confer resistance to target pathogen	
		1. Transfer of PR proteins	Tobacco plants expressing Chitinase and glucanase shows resistance against *Rhizoctonia solani*
		2. Antiviral protein (interfere with viral replication)	Tobacco and potato against PVX, Potato against PLRV and PVY
		3. Expression of bacterial lysozymes	Cecropins T4 lysozyme expressing transgenic potato against Black leg of potato
		4. Virus Coat protein	Squash plant Coat protein gene resistant against CMV, WMV2 and ZYMV.
		5. Replicase gene	Potato resistant against PVX and PVY.
		6. Movement protein gene	BMV movement protein expressed in tobacco shows resistance to TMV.
		7. Nuclear inclusion protein	Cassava against *Cassava mosaic virus*
		8. Antisense RNA	Tomato expressing anti sense RNA of TYLCV shows resistance against TYLCV.
		9. Defective interfering viral nucleic acid	*N. benthamiana* expressing DI RNA of *Tobacco bushy stunt virus* (TBSV) shows resistance against TBSV.
		10. Post transcriptional gene silencing	Transgenic papaya against PRSV

1.3 Plant Tissue Culture

Gottlieb Haberlandt (1902) proposed the theory of tissue culture. Tissues excised from different parts of plant are used as explants and grown under aseptic conditions in specific nutrient media.

Table 9.2: Various Plants Grown Resistant to Plant Pathogens by Plant Tissue Culture (Srivasatava *et al.*, 1999)

Type of Tissue Culture	*Resistance Ceveloped*
Callus culture	Tobacco regenerated from callus culture of TMV infected plant shows resistance to black shank of tobacco (*Phytophthora nicotianae*). Callus culture of maize having T-cytoplasm are resistant to *H. maydis* Callus culture of Nangeng 34 variety of rice resistant to *X. oryzae pv. oryzae.* Callus culture of Peach resistant to *X. campestris pv. pruni.*
Protoplast culture	*Nicotiana sanderae* + *N. debneyi*: Resistant to *Peronospora tabacina* *Solanum brevidens* + *S. tuberosum* = Resistant to *Potato leaf roll virus* (PLRV)
Floral mersitem	Cauliflower against *Cauliflower mosaic virus* (CaMV)

Type of Tissue Culture	Resistance Ceveloped
Meristem tip culture	Dahlia against *Dahlia mosaic virus* Potato resistant to PVX, PVY, PLRV Banana resistant to *Banana bunchy top virus* (BBTV) Chilli resistant to Chilli leaf curl virus
Meristem tip culture coupled with low temperature (5-15°C)	*Dioscorea opposita* against *Yam mosaic virus* Potato resistant *to Potato spindle tuber viriod* Argyranthemum resistant to *Chrysanthemum stunt viriod*
Meristem tip culture coupled with chemotherapy	Using antiviral chemicals like ribavirin, thiouracil, azguaninev, virizole, (RS)-9-(2, 3-dihydroxypropyl) *etc.* Kinnow resistant to *Indian citrus ringspot virus.* Potato resistant to PVX, PVY, PVS and PLRV.
Meristem tip culture coupled with Electrotherapy	Clones of potato free of PVX, PVY

1.4 Marker Assisted Selection (MAS)

Desired disease resistant genes are identified in related or unrelated lines with the help of closely linked markers. It is more efficient method of breeding as compared to conventional breeding, saves the time for phenotypic screening, helps to avoid transfer of deleterious genes, recessive and mutant genes can be selected, allows pyramiding of major disease resistance genes, genes responsible for resistance can be identified in seedling stage and helps in mapping of quantitative trait loci (QTL).

Table 9.3: Examples of Crop Varieties Raised by MAS Desistance Resistance (Ragimekula, 2013)

Crop	Trait	Genes Incorporated
Rice (PR 106)	BLB resistance	Xa5, Xa13 and Xa 21
Rice (IR 50)	Blast and BLB resistant	Piz 5, Xa 21
Rice (Co 39)	Blast	Piz 1, Piz 5 and Pita
Wheat	Brown Rust resistance	Lr 41, Lr 42 and Lr 43
Wheat (Yang 158)	Powdery mildew	Pm 2 + Pm 4a, Pm 2 + Pm 21
Wheat	Stripe rust	Yr 5 and Yr 15
Soyabean	Soyabean mosaic virus	Rsv 1, Rsv 3 and Rsv 4

1.5 Cross Protection

Plants tolerant to viral attack are made by utilizing the phenomenon of cross protection. Crop plants inoculated with an attenuated form of target virus (protecting virus) in early stages helps to combat with the challenging virus and protects the plants from viral infection. The attenuated form of virus saturates the site of replication, hinders the process of decapsidation and replication of the challenging virus, encapsidates the nascent RNA strands of the challenging virus and prevents the movement and spread of the challenging virus.

Table 9.4: List of Resistant Varieties Developed by Resistance Breeding

Crops	*Resistant varieties*	*Trait*
Rice	Improved Samba Masuri	Bacterial Leaf blight of Rice (MAS)
	Pusa 1592, Punjab Basmati-3, IR 28, Ratna, Sathi, IR 24, Ambika	Bacterial Leaf blight of Rice
	Pusa 1609, IR 22, Guj 1, IR 28, Gurjari, GR 101, Kamini	Blast of rice
	Improved DRR17B	Blast and Bacterial leaf blight
	IR 28, PUSA 33	Rice Tungro Virus
	Navagam -19, Orissa T 141	Brown leaf spot
Tobacco	Prabhat, Jayasri, Godavari Spl., MRGTH-1	TMV
	Swarna	Resistant t hectare in o Powdery mildew
	BSR 1	Black Shank
Wheat	HD 3043	Stripe rust and leaf rust
	HI 1563, Pusa Navgiri Pusa Vishesh	All 3 wheat rust
	GW 273, GW 322, HI 1500, PBW 343, UP 2338, Raj 4120 (Wheat)	Resistant to UG 99 race of PGT
	Urja	Brown rust
Pepper	IISR Sakthi	*Phytophthora capsici*
Turmeric	Pant Paatabh	*Rhizome rot*
	IISR Kedaram	Leaf blotch
Sugarcane	Shyama (Co 94008), Krishna (COM 88121)	Smut
	Sweta (CoS 94270, Rasbhari, Rajbhog, Gandak, Co Pant 90223	Red rot of sugarcane
Tomato	Arka Ananya, Kashi Vishesh	*Tomato leaf curl virus*
	Arka Abhijit, Arka Abha (BWR-1), Arka Alok (BWR-5), Shakti, BRH-1, BRH-2	Bacterial wilt of tomato
Brinjal	Arka Neelkanth, Arka Nidhi, Kt4, Hisar Shyamal, Pusa Purple Cluster	Bacterial wilt of brinjal
	Pant Samrat, Pant Brinjal Hybrid 1	Phomopsis blight
Maize	Deccan Hybrid Double Cross	Downey mildew
	Hi-strach	Leaf blight and
	Paras	Stalk rot complex
	Tarun, Kanchan, Shweta	Brown stripe downey mildew and Pythium stalk rot
Ground nut	Kadiri 3, 1 C GS-11	Bud necrosis
Red gram	Maruthi ICP 8863, Lakshmi ICPL85063, MRG 66, PRG 100	Fusarium Wilt

Crops	*Resistant varieties*	*Trait*
Potato	Kufri Ashoka, Kufri Jyoti, Kufri Jawahar	Late blight of potato
	Kufri Pukhraj, Kufri Pushkar	Early blight of potato
	Kufri Kanchan	Wart and blight
Cabbage	Pusa Drum Head	Black leg
	Pusa Mukta	Black rot
Cauliflower	Pusa Shubra, Pusa Snowball K-1	Black rot
	Pusa Hybrid 2, Pusa Kartik Sankar	Downey mildew
Pea	Pusa Pragati, Pant Sabji Matar-5,Azad P2, Azad P4	Powdery mildew
Watermelon	Arka Manik	Triple resistance to Powdery mildew, Downey mildew and Anthracnose
Chickpea	Bharati (ICCV-10)	*Fusarium* wilt and Dry root rot
	Sweta(ICCV-2)	Botrytis grey mould and wilt
	Vijay (Phule G-81-1-1), Vardan (GNG-663), GPF-2, KWR-108	Fusarium wilt
	Samrat (GNG-469)	Ascochyta blight of chickpea

2. Physical Method to Control Plant Pathogens

Various physical methods that include crop protection measures are presented in Table 9.5.

Table 9.5: Physical Methods of Plant Disease Control

Sl.No.	*Method*	*Disease*	*Causal Organism*	*Temperature*
1.	Hot water treatment	Red rot	*Colletotrichum falcatum*	54°C for 8 h
		Loose smut	*Ustilago tritici*	52°C for 11 min
		Leaf Scald	*Xanthomonas albineans*	Soaking in cold water for 1 day and then treating the cutting at 50°C for 2-3 h
		Ratoon stunting	*Clavibacter xyli*	50°C for 3h
		Black rot	*Xanthomonas campestris* pv. *campestris*	50°C for 2-3 h
		Potato	PVX, PVY, PVA, PVS	20 min treatment each day on 3 successive days at 52°C
2.	Flooding	Panama wilt	*Fusarium oxysporum* f.sp. *cubense*	
3.	Radiation		*Alternaria,Botrytis, Stemphylium*	
5.	Hot air treatment	Red rot	*Colletotrichum falcatum*	54°C for 8 h

3. Biological Control

Biological control is the reduction of inoculum density or disease producing activities of a pathogen or parasite in its active or dormant state, by one or more organisms, accomplished naturally or through manipulation of the environment, host or antagonist, or by mass introduction of one or more antagonists" (Baker and Cook, 1974).

Biological control is a sustainable approach for plant disease management as the biocontrol agents are self perpetuating and characterized with high host specificity, occurs naturally, easy to manufacture and multiply, cost effective to apply, provides protection to crops throughout the season, safe to handle, possess no residual threat, safe to beneficial micro flora, reduces the risk of resistance development, improves crop health and yield and stable than chemicals.

3.1 Suppressive Soils

Soils in relation to soil borne diseases can be classified in to two categories – conducive soils and suppressive soils. Soils where soil borne diseases take severe form are conducive soils. Whereas, those soils where the same soil borne pathogens loss their ability or have retarded ability to cause disease to the same host crops are termed to be suppressive soils. According to Cook and Baker (1983) suppressive soils are those in which the establishment of the pathogen is hindered, little or no damage is observed and the disease severity gradually reduces despite of the presence of the pathogen in soil. Antibiosis, competition, parasitism and predation are the major mechanism laying behind soil suppressiveness.

Table 9.6: Various Methods to Prepare Suppressive Soils (Bonanomi *et al.*, 2007)

Sl.No.	*Method*	*Pathogen/diseased suppressed*
1.	Soil amendment with green manure or compost	*Pythium, Phytophthora Fusarium, Sclerotinia Verticillium dahlia*
2.	Micronutruient application (Dordas *et al.*, 2008)	
	Boron (B)	Wilt of tomato
	Zinc (Zn)	Take all disease of wheat
	Manganese (Mn)	*Rhizoctonia solani* (Root rot)
	Calcium (Ca)	Take all disease of wheat
	Copper (Cu)	Soft rot of tomato and potato (*Erwinia*) Powdery mildew of wheat (*Blumeria graminis*)
3.	Crop rotation	1. Wheat-rice cropping sequence reduces Flag smut, 2. Wheat – oat cropping sequence reduces Take all disease of wheat. 3. Sunflower – Sorghum/miaze sequence reduces *Sclerotinia* sp. stalk rot
4.	Monocropping	Monocropping of wheat reduces Take all disease of wheat.
5.	Soil solarisation	*Verticillium, Fusarium, Phytophthora cinnamoni, Bipolaris sorokiniana, Rhizoctonia solani, Pythium ultimum*
6.	Biofumigation with *Brassica* spp.	*Aphanomyces, Fusarium, Gaumannomyces, Phytophthora, Pythium* and *Rhizoctonia*

Suppressive soils can be futher sub-divided into natural and induced. Natural soil suppressiveness is independent of cropping history and depends on the physical properties of soil. Soil can be induced to become suppressive by identification, multiplication and introduction of antagonists in previously conducive soils. Suppressive soils can be managed by addition of suppressive soil to conducive soil,continuous monocropping, enrichment of soil with root colonizing microflora, undertaking good agronomic practices, crop rotation, biofumigation, soil solarisation, residue management, and organic amendment of soil.

3.2 Biocontrol Agents

Biocontrol agents are microorganisms which reduces the population density of plant pathogens or reduce their ability to cause disease by direct method like antagonism (parasitism, antibiosis and competition) and indirect method.

3.2.1 Antagonism

The relation between two organisms where presence of one is detrimental to the other is known as antibiosis. Antagonism can be due to parasitism, competition or antibiosis.

3.2.1.1 Parasitism

In this phenomenon one fungus attacks other plant pathogenic fungus and inhibits the growth of the latter. Necrotrophic parasites kill the host pathogen by penetrating the host pathogen cell and degrading the cell wall, cell cytoplasm and lysing the cell. While biotrophic parasites do not kill the host pathogen but draws nutrients from the latter and suppress its growth and pathogenic ability.

Table 9.7: Examples of Biocontrol Agents Parasitizing Plant Pathogens (Sharma *et al.*, 2013)

Biocontrol Agent	*Target Pathogen*
Trichoderma viridae	*Rhizoctonia solani, Macrophomina* sp., *Fusarium* sp.
Trichoderma harzianum	*Sclerotium* sp. and *Rhizoctonia solani*
Ampelomyces quisqualis	Powdery mildew
Pichia guilliermondii	*Botrytis, Penicillium*
Pythium oligodandrum	*Pythium ultimum*
Coniothyrium minitants	*Sclerotinia sclerotiorum, Botrytis, Penicillium etc.*
Laetisaria arvalis	Sclerotia producing fungi
Talaromyces flavus	*Verticillium* sp., *Rhizoctonia solani* and *Sclerotinia sclerotiorum etc.*

3.2.1.2 Competition

Common source of food, shelter and nutrients results in competition for survival among two or more organisms. Competition may prevent the entry of the pathogen at the infection site or retard the reproduction of the pathogen.

Table 9.8: Examples of Biocontrol Agents Parasitizing Plant Pathogens

Biocontrol Agent	Target Pathogen
Peniophora gigantean	*Heterobasidium annosum* causing red rot of pine
Erwinia herbicola	*Erwinia amylovora* causing fire blight of apple and pear
Idriella bolleyi	*Gaeumannomyces graminis* f.sp. *tritici*
GL 20 of *Pseudomanas fluorescens* (Produce siderophores)	*Fusarium solani*

3.2.1.3 Antibiosis

Antibiosis includes suppression of growth of target pathogen due to secretions of low molecular weight toxic compounds or antibiotics by bio-control agents.

Table 9.9: Examples of Antibiotics Produced by Biocontrol Agents (Junaid *et al.*, 2013)

Antibiotic	Biocontrol Agent	Target Pathogen
Trichodermin	*T. viridae/T. harzianum*	*R. solani, Sclerotinia sclerotiorum*
Acetaldehyde	*T. viridae*	*R. solani, Macrophomina paseolina*
Epirodoxin	*Epicoccum purpurascens*	*Colletotrichum capsici,. lagenarium, Macrophomina paseolina*
Laetisaric acid	*Laetisaria arvalis*	*R. solani* and *Pythium ultimum*
Phenazines	*Pseudomonas fluorescens* 2-79 and *30-84*	*Gaeumannnomyces graminis* f.sp. *tritici*
Zwittermycin A	*Bacillus cereus* UW85	*Pythium aphanidermatum*
Iturin A	*Bacillus subtilis* QTS713	*R. solani and Botrytis cinerea*
Mycosubtilin	*B. subtilis* BBG100	*Pythium aphanidermatum*
Pyrrolnitrin	*Pseudomonas fluorescens* Pf-5	*Gaeumannnomyces graminis* f.sp. *tritici*
Agrocin -84	*Agrobacterium radiobacter*	*Agrobacterium tumifaciens*
2,4- diacetylphloro-glucinol	*Pseudomonas fluorescens* F113	*Pythium* spp.
Bacillomycin D	*B. subtilis* AU195	*Aspergillus flavus*

3.2.2 Indirect Method

The indirect method of bio control agent activity includes hypovirulence, cross-protection, induced resistance and growth promotion.

3.2.2.1 Hypovirulence

Reduced virulence of a pathogen strain caused due to infection by transmissible ds RNA.

3.2.2.2 Cross Resistance

Mild strain of a virus (protecting virus) helps to combat with the challenging virulent virus by various mechanisms including decapsidation, inhibition of replication, transcription, translation and post transcriptional gene silencing of the challenging virus.

Table 9.10: Examples of Hypovirulence

Associated RNA/DNA	*Hypovirulent Strain*	*Disease Controlled*
CHV 1	*Cryphonectria parasitica*	Chestnut blight
MOv1 AND MOV2	*Magnaporthe grisea*	Blast of rice
Rhizoctonia virus N2	*Rhizoctonia solani*	Sheath blight of rice
BVF and BVX	*Botrytis cinerea*	Grey mould

Table 9.11: Examples of Cross Resistance Used in Plant Pathogen Control

Virus	*Crops Protected*
Citrus tristeza virus (CTV)	Citrus
Zucchini yellow mosaic virus (ZYMV)	Squash, cucumber *etc.*
Papaya ring spot virus (PRSV)	Papaya, squash
Soyabean mosaic virus	Soyabean

3.2.2.3 Induced Resistance

Resistance to target pathogens are induced by various soil microflora and chemicals.

3.2.2.3.1 Plant Growth Promoting Rhizobacteria

These bacteria that promote growth rate by colonizing the plant roots and compete with other organisms are known as Plant growth promoting rhizobacterium.

Table 9.12: Examples of PGPR Controlling Plant Pathogens

Sl.No.	*Mechanism*	*PGPR/Chemicals*	*Target Pathogen*
1.	Production of antibiotic and lytic enzymes	*HCN, Phenazines,* 2,4-diacetyl phloroglucinol, Pyoluteorin and Pyrrolnitrin	*Thielaviopsis basicola, Gaeumannnomyces graminis* f.sp. *tritici, Pythium* spp.
		Chitinase, cellulose,lipases, 1,3-glucanase and proteases	Squash, cucumber *etc.*
2.	Competition	*Pseudomonas fluorescens* PCL1751	*Fusarium fluorescens* f.sp. *radicis-lycopersici*
3.	Induced Systemic resistance	Induces Jasmonic acid and ethylene signaling (*Pseudomonas fluorescens*)	*Gaeumannnomyces graminis* f.sp. *tritici*, Sheath blight disease

3.2.2.3.2 Plant Defense Activators

Many synthetic and natural compounds when applied as foliar spray or soil drench stimulate the signal transduction pathway mediated by salicylic acid and also up-regulates various defence genes *viz.* pathogenesis-related (PR) proteins, phenolics, phytoalexins, and terpenoids *etc.*

Table 9.13: Examples of Plant Defence Activators Used to Control Plant Pathogens (Sreeja, 2014)

Sl.No.	*Biocontrol product*	*Source*	*Effective against pathogens*
1.	Actigurad	Benzothiadiazole	Wide
2.	Blockade	Synthetic compound	Downey mildew
3.	Actinovate	*Streptomyces lydicus*	*Rhizoctonia solani, Macrophomina phaseolina*
4	AQ 10 biofungicide	*Ampelomyces quisqualis* M10	Powdery mildew
5	Aspire	*Candida olephila-182*	*Botrytis* spp., *Penicillium* spp.
6	Messenger	*Erwinia amylovora* haipin protein	Many pathogens
7	Acebenzolar-S-methyl (ASM)	Benzothiadiazole derivative	Many pathogens
8	Yield Shield	*Bacillus pumilis* GB 34	Root rot pathogens
9	INA	2,6-dichloroisonicotinic acid	
10	BABA	*β-aminobutyric acid*	Late blight of tomato, Downy mildew of grape vine and *Phytophthora* blight of pepper
11	Oryzemate	Probenazole	Blast of rice
12	Chitosan		
13	Oligogalacturonides	Plant cell pectin	
14	Hairpin protein	Encoded by hrp genes	Many pathogens

Table 9.14: List of Commercially Available Biocontrol Agents (Junaid *et al.*, 2013)

Biocontrol Agent	*Product*	*Target Pathogen*
Bacillus subtilis strain GB03	Companion	*Pythium, Phytophthora, Fiusarium* and *Rhizoctonia*
Bacillus subtilis strain MB1600	Histick N/T	*Fusarium, Aspergillus, Rhizoctonia*
Bacillus subtilis GB03	Kodiak	*Rhizoctonia solani and Fusarium*
Pseudomonas fluorescens	Dagger G	*Pythium and Rhizoctonia*
Bacillus subtilis strain GB03	Kodak	*Fusarium, Aspergillus, Rhizoctonia* and *Alternaria*
Burkholderia cepacia	Deny	*Pythium, Fusarium, Rhizoctonia*
Trichoderma harzianum	F- Stop	*Pythium* and *Rhizoctonia*
Gliocladium virens GL-21	SoilGard	*Pythium* and *Rhizoctonia*
Ampelomyces quisqualis isolate M-10	*AQ 10*	Powdery mildew
Trichoderma harzianum T-22	*Root shield*	*Rhizoctonia solani* and *Fusarium* sp.
Agrobacterium radiobacter strain 84	Galltrol	*Agrobacterium tumifaciens*
Agrobacterium radiobacter strain K 1026	*NOgall*	*Agrobacterium tumifaciens*
Bio-save 10 LP, 110	*Pseudomonas fluorescens*	*Botrytis, Penicillium, Aspergillus*
Aspergillus niger AN-27	Kalisena	*Rhizoctonia solani*
Candida oleophila I-182	Aspire	*Botrytis, Penicillium*

Biocontrol Agent	Product	Target Pathogen
Fusarium oxysporum (*non pathogenic*)	Biotrox C, Fuscaclean	*Fusarium oxysporum*
Coniothyrium minitants	Contans WG, Intercept WG	*Scletorinia sclerotiorum*
Pythium oligandrum	Polygandron	*Pythium ultimum*
Gliocladium catenulatum	Primastop	Root rots and wilt
Trichoderma harzianum, T. polysporum	BINAB T	Wood decay fungi
Trichoderma harzianum, T. viridae	Promote	*Pythium, Fusarium and Rhizoctonia.*
Phlebia gigantean	RotStop	*Heterobasidion annosus*
Trichoderma harzianum	Trichodex	*Colletotrichum, Monilinia*
Trichoderma harzianum	Trichopel	*Armillaria, Fusarium,*
Trichoderma harzianum and *T. viridae*	Trichojet	*Nectria, Phytophthora, Pythium* and *Rhizoctonia*

4. Use of Botanicals

Plant extracts are eco-friendly novel chemo-therapeutants that possess antimicrobial property confers broad spectrum resistance and in the same time is easily biodegradable and come with a low expense to farmers. Use of botanicals reduces the risk of resistance development in target pathogen, safer to non target beneficial micro flora, quickly biodegradable, cost effective and is a component of Integrated plant disease management (IPM).

Table 9.15: List of Plant Extracts Used to Control Plant Diseases (Gurjar *et al.*, 2012)

Sl.No	Plant	Compound	Resistance against
1	Neem (*Azadirachta indica*)	Azadirachtin, nimbin, nimbidin *etc.*	Sheath rot of rice (*Sarocladium oryzae*) Powdery mildew of rose (*Sphaerotheca pannosa*) Black spot of roses (*Diplocarpon rosae*)
2	Garlic (*Aliium sativum*)	Allicin	Against loose smut of wheat, *Alternaria solani, Aspergillus niger* and *A. flavus.*
3	Onion (*Allium cepa*)	Allicin	Phytophthora leaf blight of tomato, Blast of rice,
4	Apple (*Malus pumila*)	Phloretin	Apple Scab incited by *Venturia inaequalis*
5	Ashwagandha (*Withaia somnifera*)	Withafarin A	Fusarium wilt, *Aspergillus flavus etc.*
6	Thyme (*Thymus vulgaris*)	Caffeic acid	*Penicillium digitatum* (Green mould of citrus) and various virus and bacteria
7	Turmeric (*Curcuma longa*)	Curcumin	*Alternaria solani* in tomato, *Phytophthora infestans, Pyricularia oryzae*
8	Black pepper (*Piper nigrum*)	Piperine	Fusraium wilt

Sl.No	*Plant*	*Preparations and Volatile Oil*	*Resistance against Pathogen/Disease*
1	Garlic (*Aliium sativum*)	Ethanol extracts	Curvularia lunata
2	Neem (*Azadirachta indica*)	NSKE	*Alternaria alternate,* Powdery mildew of pea (*Uncinula necator*)
3	Hena (*Lowsonia inermis*)	Crude extract	Helminthosporium oryzae
4	Perslane (*Portulaca olaracea*)	Leaf extract	*Helminthosporium maydis*
5	Tulsi (*Ocimum tenuiflorum*)	Leaf extracts	Blast of rice (*Pyricularia oryzae*) and brown spot of rice (*Helminthosporium oryzae*)
6	Mint (*Mentha longifolia*)	Leaf extract	Brown spot of rice
7	Datura (*Datura stramonium*)	Aqueous extracts	*Aspergillus* spp.
8	Rapeseed (*Brassica napus*)	Crude extract	Bacterial diseases of onion
9	Madar (*Calotropis procera*)	Crude extracts	Tikka disease of groundnut
10	Spear mint (*Mentha spicata*)		*R. solani, Sclerotinia sclerotiorum*
9	Lemon grass (*Cymbopogon* sp.	Volatile oil	Black mould of onion bulbs (*Aspergillus niger*)
10	Dawn red wood (*Metasequoia glyptostroboides*)	Essential oil	Fusarium wilt, *Phytophthora capsici, Colletotrichum capsici* and *Sclerotinia sclerotiorum* and *Rhizoctonia solani*
11	Basil (*Osimum basilicum*)	Essential oil	Grey mould of grapes (*Botrytis cinerea*)
12	Thyme (*Thymus vulgaris*)	Essential oil	*Alternaria alternate*
13	*Hyssopus officinalis*	Hyssop oil	Blast (*Pyriculatia oryzae*) and seedling blight of oat(*Pyrenophora avenae*)
14	Nettle (*Urtica dioica*)	Nettle oil	*Alternaria solani*
15	Eucalyptus	Oil	*Rhizoctonia solani, Penicillium digitatum, Alternaria brassicae* and *Albugo candida etc.*

5. Fungicide Resistance Management

A stable, heritable trait obtained through evolutionary processes that result in a reduction in sensitivity to a fungicide by an individual fungus is known as fungicide resistance. The agricultural industry is facing problem due to fungicide resistance since 45 years. Problem first appeared with registration and widespread use of the systemic fungicide benomyl (Benlate) in the early 1970's. First fungi reported to show fungicide resistance was *Podosphaera xanthii.*

Table 9.16: Fungicides Proved to be Ineffective Due to Fungicide Resistance (FRAC)

Date First Observed	*Fungicide/ Fungicide Class*	*Years of Commercial Use*	*Main Crop Diseases and Pathogens Affected*
1960	Aromatic hydrocarbons	20	*Penicillium* sp.
1964	Organo-mercurials (PMA)	40	Cereal leaf spot and stripe-*Pyrenophora* sp.
1969	Dodine	10	*Venturia inaequalis*
1970	Benzimidazoles (Benomyl)	2	Many target pathogens

Date First Observed	Fungicide/ Fungicide Class	Years of Commercial Use	Main Crop Diseases and Pathogens Affected
1971	2-Amino-pyrimidines (Ethirimol)	2	Cucumber and barley powdery mildews- *Sphaerotheca fuliginea* and *Blumeria graminis*
1971	Kasugamycin	6	*Magnaporthe grisea*
1976	Phosphorothiolates	9	*Magnaporthe grisea*
1980	Phenylamides (Metalaxyl)	2	*Phytophthora infestans* and *Plasmopara viticola*
1982	Dicarboxamides (Iprdione)	5	*Botrytis cinerea*
1982	Sterol Demethylation Inhibitors (Triadimefon)	7	Cucumber and barley powdery mildews- *Sphaerotheca fuliginea* and *Blumeria graminis*
1985	Carboxanilides	15	*Ustilago nuda*
1998	Quinone oxide inhibitors (Qols, Strobilurins) (Azoxystrobin)	2	Many target diseases and pathogens (Cucumber Powdery mildew and Downey mildew)
2002	Melanin Biosynthesis Inhibitors	2	Magnaporthe grisea

Table 9.17: Reduction in Resistance Development by Mixed or Alternate Application of "At risk fungicide" with "Companion fungicide" (FRAC)

Fungicide Combination	Target Pathogen	Status of at Risk Fungicide
Phenylamide + Mancozeb	*Phytophthora infestans*	Decreased the build-up of phenylamide resistance
Fosetylaluminium (Folpet) + Mancozeb	*Plasmopara viticola*	Selection for QoI resistance was delayed.
Benomyl+ Organo tin fungicide	*Cercospora beticola*	Delayed development of Benomyl resistance.
Tridimenol + Sulphur/Dinocap	*Uncinula necator*	Decreased resistance development against Tridimenol
Azoxystrobin + Epoxiconazole	*Mycosphaerella graminicola*	Reduced build up of resistance to QoI fungicides.
Triazoles+ Morpholines	*Blumeria graminis* f.sp. *tritici*	Delayed resistance to both of the fungicide.

Table 9.18: Recommended Fungicides for some Diseases to Overcome Risk of Fungicide Resistance

Fungicides	Diseases
Rice Blast	Tilt 25 EC (Propiconazole) @ 200 ml/Indofil Z- 78
Sheath blight of Rice	Lusture/Folicur/Tilt/Moncern/Nativo
False Smut of Rice	1st spray with Kocide and 2nd spray with Tilt.
Foot Rot of Rice	Tilt

Fungicides	*Diseases*
Yellow rust of Wheat	Tilt/Shine/Folicur
Downey mildew of Maize	Indofil M- 45
Loose Smut	Vitavax/Raxil
Karnal Bunt of Wheat	Tilt/Folicur

REFERENCES

A. Sharma, V.D. Diwevidi, S. Singh, K K Pawar, M Jerman, L.B. Singh, S Singh and D Srivastawa (2013). Biological Control and its Important in Agriculture. *International Journal of Biotechnology and Bioengineering Research* **4** (3): 175-180.

C Dordas (2008). Role of nutrients in controlling plant diseases in sustainable agriculture. In book-A *review Agronomy for Sustainable Development*. Pp-33–46

G. Bonanomi, V. Antignani, C. Pane and F. Scala (2007). Suppression of soilborne fungal diseases with organic amendments. *Journal of Plant Pathology* **89** (3): 311-324

J. M. Junaid, N. A. Dar, T. A. Bhat, A. H. Bhat and M. A. Bhat (2013). Commercial Biocontrol Agents and Their Mechanism of Action in the Management of Plant Pathogens. *International Journal of Modern Plant and Animal Sciences* **1**(2): 39-57

M. S. Gurjar, S. Ali, M. Akhtar and K. S. Singh (2012). Efficacy of plant extracts in plant disease management. *Agricultural Sciences*. **3** (3): 425-433.

N. Ragimekula, N. N. Varadarajula, S. P. Mallapuram, G. Gangimeni, R. K. Reddy, H. R. Kondreddy (2013). Marker assisted selection in disease resistance breeding. *Journal of Plant Breeding and Genetics* 1(2): 90-109

P. S. Srivastava, Muhammad Iqbal, M. Haseeb Mughal (1999). Role of Tissue Culture in Plant Disease Control. In book: *Biotechnological Approaches in Biocontrol of Plant Pathogens* pp.197-217.

S.J Sreeja (2014). Synthetic Plant Activators for Crop Disease Management – A Review. International Journal of Thesis Projects and Dissertations. **2**(1): 19-28.

V Kumar and S Chandel (2016). Mycoviruses and their role in biological control of plant diseases. *International journal of Plant Sciences*. **11** (2): 375-382

Conservation Agriculture Technologies *Pages* **111–119**
Editor: **Rahul Kumar, Abhinandan Singh and Pankaj Kumar Ojha**
Published by: **BIOTECH BOOKS, NEW DELHI**

10 Efficient Conservation Techniques for Natural Enemies

I. Yimjenjang Longkumer

Department of Entomology, Central Agricultural University, Imphal – 795 004

Due to the consistency in insecticide application and removal of the crop after the season, the life expectancy of the Natural enemies drastically decreases with the following crop season. The effects of pesticide have lethal and sub-lethal adverse effect on the natural enemies. Conservation is the techniques that involve the protection to the natural enemies from the adverse effect of the cultural practices, pesticide application and environmental hazards. The strategies involved in the conservation have played a major impact in the efficiency of the natural enemies in controlling the insect pest in every crop season. Collection of the natural enemies before the crops are harvested, preservation of the collected natural enemies in Greenhouse where the climate is conducive for the normal development and survival of the natural enemies during non-crop period and release of the natural enemies on the next crop season to the target crop accomplish the conservation technique.Such a conservation biological control strategy might contribute to preserve the natural biodiversity in the agricultural environment and provide natural alternatives to pesticides.

Introduction

Biocontrol is defined as the pest population regulations by the activities of Natural enemies. It employs various techniques *viz.* introduction, augmentation and conservation. In classical biological control, an NE is imported and released in a new area for regulating a specific pest(Perdikis *et al.*, 2012). In modern agricultural system, the populations of the released and new Natural enemies are deteriorated due to the complete removal of the crop after harvest which leads to the disappearance of the Natural enemies after each crop season.Conservation biological control is the protection of NEs against adverse effects ofpesticides and incompatible cultural practices and improving their efficiency via providingfood sources(Rabb *et al.*,

1976). DeBach (1974) defines conservation biological control as "manipulation of the environment to favor natural enemies, either by removing or mitigating adverse factors or by providing lacking requisites". This broad definition emphasizes the fundamental importance of conservation to all approaches of biological control. The efficacy of imported or mass reared natural enemies in many systems will likely depend on attention to conservation measures that improve the suitability of the environment into which theyare released (Gurr and Wratten, 1999). Since time immemorial biological controls of arthropod pest have been used as Natural Enemies which plays an important role in limiting potential pest populations. Conservation biological control is one of the biological control main branches(Debach and Rosen, 1991) which can be first realized by reducing the use of pesticides, application of selective pesticides, proper placement and timely pesticide applications otherwise Insects that were of little economic importance may become destructive pests. When nontoxic control method is used natural enemies are more likely to survive and reduce the populations of pests.

During off season, flowering plants within or along the edges of the field enhances the survival of NE as many crop plants have short flowering period. Preservation of natural enemies can be achieved by providing habitat and resources for natural enemies(Fiedler*et al* 2008).Mostly they remain inactive during the winter. Unless they are re-released each year, they must have a suitable environment for overwintering (Sotherton, 1984; Leather, 1993). They usually pass the winter in crop residues, other vegetation or in the soil. Ground cover of fruit orchards, winter crops (like alfa alfa and breccias), usually provides shelter for overwintering natural enemies. Adding plants or other food sources for natural enemies must be done with knowledge of the behaviour and biology of the natural enemy and the pest (Bianchi *et al.*, 2008; Landis *et al.*, 2000; Maoz *et al.*, 2011; Wäckers *et al.*, 2012). Monoculturing is the greatest threat to the environmental quality, decreasing biodiversity and even results in the pest outbreaks. Modern crops are not only monocultured in highly specialized production unitsbut even the harvest and packaging techniques are specialized. Hence the dependence of the agroecosytem with the ecological potential such as natural enemies in pest suppression will increase the sustainability of the agroecosystem. Farming systems like greenhouses, annual crop systems and other practices that end with removing the whole crop after harvesting, may give rise to elimination of biodiversity, and decreasing the population of natural enemies in the fields or in different agricultural environments (Lavandero *et al.*, 2006; Gurr *et al.*, 2012). This eventually will preserve the NE and natural biodiversity in the environment and provide an efficient step as the alternative to chemical uses.

This chapter emphasize on the efficient techniques for the conservation of natural enemies:

1. Collection of natural enemies before the end of crop season.
2. Preservation of collected natural enemies in special greenhouses during non-crop periods
3. Releasing the preserved natural enemies on target crops in the next growing season

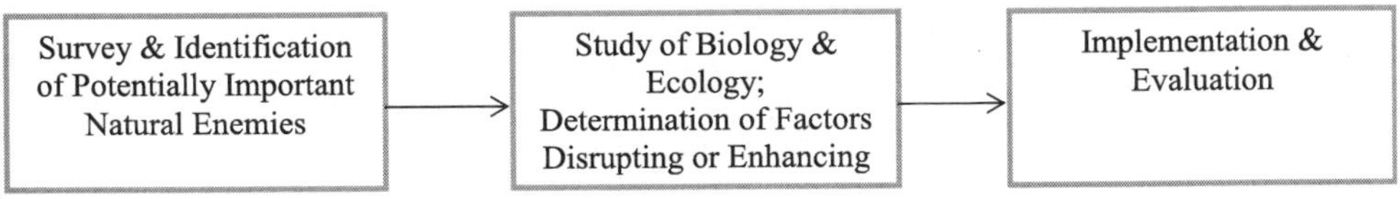

Figure 10.1: Components and Approaches of Conservation Biological Control (Naranjo, 2001)

1. Collection of Natural Enemies

Before complete removal of the crop the natural enemies are collected so as to prevent their population disappearance.At the end of the crop season, the NEs are usually in their top population densities (Perdikis *et al.*, 2011).

Table 10.1: Crops with their Natural Enemies

Crops	*Natural Enemies*	*Pests*	*Techniques*
Mulberry	*Encarsia citrina* *Anagyrus kamali* *Metaphycus* sp. *Allotropa mecrida* *Scutellista caerulea* *Chartocerus* sp. *Brevipalpus* sp. **Predators** *Orius sp.* *Coranus* sp. *Coccinella undecimpunctata* *Cydonia* sp. *Mantis religiosa*	**Parasitoids** *Panonychus ulmi* *Thrips tabaci* *Nezara viridula* *Bemisia tabaci* *Aphis gossypii* *Icerya aegyptiaca* *I. purchase* *I. seychellarum* *Ceroplastes rusci* *Coccus hesperidum* *Saissetia oleae*	**Parasitoids** Picking infested leaves containing parasitized insects **Predators** Individuals were collected by beating tree branches in a suitable cloth bag
Tomatoes	*Nesidiocoris tenuis* *Chrysoperla carnea*	*Tuta absoluta* *Phthorimaea opercullela* *Bemisia tabaci*	Sweeping net
Potatoes	*Hypera postica* *Apanteles* spp.	Phthorimaea operculella Agrotis ipsilon	Tomato or potato leaves were collected in jars and kept in the laboratory until parasitoids emergence
Mango trees	*Amblyseius* spp. *Oligonychus mangiferus* *Brevipalpus obovatus*	*Aulacaspis tubercularis* *Kilifia acuminate*	Infested small branches were collected in cloth bags and the predators were counted

Crops	*Natural Enemies*	*Pests*	*Techniques*
Pigeon pea	*C.septempunctata*	*Aphis fabae*	
	Andrallus spinidens	*Oxyrachis tarandus*	
	Rhynocoris fuscipes	*Odontotermes*	
	Componotus sp.		
	Mantis religiosa obesus		
	Nezara viridula		
	Melanoplus bivittatus		

Source: Nabil *et al.,* 2017.

High numbers of natural enemies are found during the cultivated season.In maize plants, high numbers of predators are built up after heavy infestation by the aphids. These predators can be collected for mass rearing in the greenhouse and for preservation or they can be directly transferred to the target crops in need of them. Populations of the Natural enemies are also present in high number during the end of the cropping season which will be lost after harvesting and removal of the plant. These NE can be collected and grown in greenhouse during non-crop season and then released in the next season.

2. Preservation of Natural Enemies

The collected natural enemies are grown in the Greenhouses with their required environmental conditions.Preservation of the Natural enemies is the cornerstone of Conservation of Biological control. Preservation can be applied individually or in combination. Applied practices for preservation of natural enemies in different fields are summarized in Table 10.2.

Table 10.2: Applied Practices for Preservation of Natural Enemies

Crop	Natural enemy	Pest	Practice
Sweet pepper, ornamental crops	Syrphids lacewings hoverflies Predatory mites, *Orius laevigatus* *O.majuscules, O.insidiosus*	White flies, thrips, aphids	Plants providing pollen and plantsap as food sources for naturalenemies like sweet alyssum,coriander, Ricinus communis andflowering ornamental
Cereal crops	Aphid parasitoids	Cereal aphids	Introducing non-crop plants harbouring the prey species
Tomato, sweet pepper	General mired predators, *Orius* spp., lacewings	Whitefly, leaf miners, *T. absoluta*	Planting suitable non-crop plantsnear fields that help natural enemies to migrate into fields
Cotton, tomato	*Orius* spp., lacewings, lady beetles	Aphids, thrips, leaffeeders	Induced plant responses that attract and/or retain natural enemies

Crop	Natural enemy	Pest	Practice
Wheat	*Orius* spp., lacewings, lady beetles	Aphids, thrips, leaffeeders	Mitigation of pesticide side-effects by selecting pesticides that are compatible with natural enemies
Ornamentals thrips	*Orius insidiosus*		Mixed diet of prey, or mixes of prey and non-prey food sources

Source: Nabil *et al.*, 2017.

2.1 Plants Providing Food Resources

Practices of preservation of natural enemies are many and vary according the types of natural enemies, the target pests, the plants and the ecological conditions. The effect of the food source *i.e.* pollen, nectar and plant sap to Natural enemies depends on the type of predators and parasitoids. Adults of parasitoids and gall midges can increase their longevity, flight activity and oviposition by feeding on nectar (Wäckers *et al.*, 2005). High populations can also be achieved by growing of crops like coriander. Plants that produce a lot of pollen, like *Ricinus communis,* provided more pollen to predatory mites(Ramakers and Voet, 1995). Thus, the availability of plants provided food can be a driving force in the conservation of biological control.

2.2 Introducing Non-Crop Plants

A widely applied system in different crops has been the use of monocotyledonous plants with cereal aphids that serve as alternative hosts for parasitoids of aphids that attack the dicotyledon crop (Huang *et al.*, 2011; Frank, 2010). Prey/host plants can also be established on the edges of the field to bridge non-crop periodsand contribute to the preservation of natural enemies (Arno *et al.*, 2000).

2.3 Planting Suitable Non-crop Plants Near Fields

Preservation biological control of predators can be enhanced by planting suitable non-crop plants near fields for the survival of the Natural enemies by sheltering them once the crops are harvested or removed and also to support migration into the crop. Providing overwintering shelters may enhance lacewings by providingdiapausing adults with artificial overwintering chambers in greenhouses (Thierry *et al.*, 2002). These methodsmay contribute to early establishment of natural enemies in new season in the spring.

2.4 Induced Plant Resistance

Induced plant resistance involves the direct effect to the insect by production of toxins and feeding deterrents that reduce survival, host preference, fecundity or developmental rate of pests and indirect effect includes attracting the preys/ Natural enemies. Plant volatiles help natural enemies to detect their prey/hosts in a crop whereas and it increases when crops are attack by an insect, guiding natural enemies to find their prey/hosts. Preservation of natural enemies might be enhanced in different cropsby breeding varieties that produce more volatiles and nectar.

3. Release of Natural Enemies

The techniques for the release of the NE vary with the type of the bioagents, hosts and weather conditions. In case of egg parasitoids, they are released as parasitized egg patches; larvalparasitoids are released as adults. Predators are usually released in the pupal stage. Timing,rate and frequency of release are determined according to the nature of the target pests, naturalenemies and crops. Pathogens like Entompathogenic nematodes could be applied as sprays orinjection (El-Wakeil *et al.*, 2009; Saleh *et al.*, 2014; Saleh *et al.*, 2015) Examples of cases of NE field releasing are summarized under the following table.

Table 10.3: Release Techniques Used for Various Batural Enemies in different Crops

Crop	*Natural Enemies*	*Pests*	*Techniques*
Tomatoes	Egg parasitoids *Trichogramma*	*Tuta absoluta*	Paper cardboard or strips containing about 400 parasitized eggs of *Ephistiakuehniella* ready to emerge
Cabbages	*Trichogramma*	*Pieris rapae*	Releasing *Trichogramma* to control *Pieris rapae* (**tricho card)**
Cotton	*Trichogramma*	Bollworms	Releasing *Trichogramma* in cards, each contain 1000 parasitoid for several times
Sugarcane	*Trichogramma*	*Chilo* sp	Released of 30,000–120,000 parasitoids
Maize	*Larval parasitoids* *Bracon* spp	Corn borers	Larval and pupal parasitoids are released in the pupal stage on special carriers like talc powder
Date palms	*Entomopathogenic nematodes*	Red palm weevil	Spraying EPNs around infested tree trunks
Ornamental plants	*Mealybug parasitoids* *Anagyrus kamali* and *Gyranusoidea indica*	*Maconellicoccus hirsutus*	300,000 parasitoids in parasitized individual stage were released

Egg Parasitoids

Tricho Card

The parasitisation of *Trichogramma* spp., in laboratory condition on one cc eggs of *Corcyra cephalonica*, which are uniformly spread and pasted on a card measuring 15 cm x 10 cm, is called as Tricho card. The card has 12 demarcations (stamps). About 12,000 Trichogramma adults emerge out from this card in 7-8 days after parasitisation. Trichocards have a shelf life of 2-3 days.

Egg parasitoids are commonly released inTricho cards or strips holding theparasitized eggs. Cardboard strips containing parasitized eggs in tubes are released in tomatoes for controlling *T. absoluta*. In cotton, The Trichocards are released in the field at 45 days after sowing @ 5 cards/ha (one lakh eggs). In total three releases are necessary.

Trichogramma chilonis, T. japonicum against lepidopterous insect pestin rice fields. *Trichogramma buesi* against *Pieris rapae* eggs in cabbage fields.

For sugarcane early shoot borer 6,000 parasites per week per acre area, for a period of 5 weeks, starting from 4th week of planting *i.e.*, as soon as the adult male moths of early shoot borer are noticed in the field.

Larval and pupal parasitoids are released in the pupal stage. Parasitized pupae just beforeemergence are carried on special carriers like talc powder and distributed in the target fields.

White Fly Parasitoids

Encarsia spp. or *Eretmocerus* spp. are released as parasitized pupae shortly before adultemergence. Simmons and Abd-Rabou (2005)confirmedthat inundative releases of parasitoid *Eretmocerus mundus* against *B. tabaci* into tomato andcotton fields increased parasitization rates.

Conclusion

With the appropriate conservation of Natural Enemies it can serve as an alternative to pesticide usage which often creates hostile environment for the beneficial insects as well as the Natural enemies in the cropping system. It also serves as a free control component in an agro system as they are readily available in the ecosystem which contribute in the control of the pest and ultimately enhances the yield of the cultivated crops. The techniques involves for the conservation of Natural Enemies are low input and feasible in any area. Hence, Natural Enemies can be employed as a single component in the Integrated Pest Management.

REFERENCES

Arno J, Arino J, Espanol R, Marti M and Alomar O: Conservation of Macrolophus caliginosusin greenhouses during tomato crop-free periods. IOBC/WPRS Bull 2000; 23: 241–246.

Bianchi FJJA, Wäckers FL: Effects of flower attractiveness and nectar availabilityin field margins on biological control by parasitoids. *Biol Control* 2008; 46: 400–408.

Debach P, Rosen D: Biological control by natural enemies (2nd edn). Cambridge UnivPress, Cambridge, UK, 1991, p. 440.

DeBach, P., 1974. Biological Control byNatural Enemies. Cambridge UniversityPress, London, 323 pp.

El-Wakeil NE, Hussein MA: Field performance of entomopathogenic nematodes andan egg parasitoid for suppression of corn borers in Egypt. *Archiv Phytopathol Plant Prot* 2009; 42: 228–237.

Fiedler AK, Landis DA, and Wratten SD: Maximizing ecosystem services fromconservation biological control: The role of habitat management. Bio Control 2008; 45: 254–271.

Frank SD. Biological control of arthropod pests using banker plant systems: Past

Gurr GM, Wratten SD, Snyder WE and Read DMY: Biodiversity and insect pests. Key issues for sustainable management. Wiley, West Sussex, UK, 2012.

Gurr, G.M and Wratten, S.D., 1999. Integrated biological control': a proposal for enhancing success in biological control. *Int. J. Pest Manage.* 45, 81–84.

Huang NX, Enkegaard A, Osborne LS, Ramakers PMJ, Messelink GJ, Pijnakker J, Murphy G: The banker plant method in biological control. *Crit Rev Plant Sci* 2011; 30: 259–278.

Landis DA, Wratten SD, Gurr GM: Habitat management to conserve natural enemiesof arthropod pests in agriculture. *Annu Rev Entomol* 2000; 45: 175–201.

Lavandero IB, Wratten SD, and Didham RK, Gurr G: Increasing floral diversity for selective enhancement of biological control agents: A double-edged sword? *Basic Appl Ecol* 2006; 7: 236–243.

Leather SR: Overwintering in six arable aphid pests: A review with particular relevanceto pest management. *J Appl Entomol* 1993; 116: 217–233.

Maoz Y, Gal S, Argov Y, Coll M, Palevsky E: Biocontrol of persea mite, *Oligonychus perseae*, with an exotic spider mite predator and an indigenous pollen feeder. *BiolControl* 2011; 59: 147–157.

NaranjoSteven E.Crop Protection (2001).Conservation and evaluation of natural enemies in IPM systemsfor *Bemisia tabaci. Crop Protection* 20: 835–852

Perdikis D, Fantinou A, and Lykouressis DP: Enhancing pest control in annual crops by conservation of predatory Heteroptera. *Biol Control* 2011; 59: 13–21.

Progress and future directions. *Biol Control* 2010; 52: 8–16.

Rabb RL, Stinner RE, van den Bosch R: Conservation and augmentation of naturalenemies. In: Huffaker CB, Messenger PS (Eds.), Theory and Practice of BiologicalControl. Academic Press, New York, NY, 1976; pp. 233–254.

Ramakers PMJ and Voet SJP: Use of castor bean, Ricinus communis, for the introduction ofthe thrips predator *Amblyseius degenerans* on glasshouse-grown sweet peppers. *MedFac Landbouww Rijksuniv Gent* 1995; 60: 885–891.

Saleh MME, Hussein MA, Hafez GA, Hussein MA, Salem HA, Metwally HMS: Foliarapplication of entomopathogenic nematodes for controlling Spodoptera littoralis andAgrotis ipsilon on corn (Zea mays) plants. *Acta Advances in Agric Sci* 2014; 3: 51–61.

Saleh MME, Hussien MA, Metwally HMS, Ebadah IM: Comparative study of qualitytraits of entomopathogenic nematodes before and after passing through certain insecthosts. *Egypt J Bio Pest Cont.* 2015; 25: 237–243.

Simmons AM, Abd-Rabou S: Parasitism of Bemisia tabaci after multiple releases ofEncarsia sophia in three vegetable crops. *J Agric Urban Entomol* 2005; 22: 73–77.

Sotherton NW: The distribution and abundance of predatory arthropods overwinteringon farmland. *Ann Appl Biol* 1984; 105: 423–429.

Thierry D, Rat-Morris E, Caldumbide C: Selective attractivity of artificial overwinteringchambers for the common green lacewing species of the Chrysoperla carnea complex inWestern Europe (Neuroptera: Chrysopidae). *Acta Zool Acad Sci Hung* 2002; 48: 351–357.

Wäckers FL, van Rijn PCJ, Bruin J (Eds.): Plant-provided food for carnivorous insects: A protective mutualism and its applications. Cambridge Univ Press, Cambridge, UK, 2005.

Wäckers FL, van Rijn PCJ: Pick and mix: Selecting flowering plants to meet therequirements of target biological control insects. *Biodiversity and Insect Pests.* 2012; pp. 139–165.

Conservation Agriculture Technologies *Pages* **121–131**
Editor: **Rahul Kumar, Abhinandan Singh and Pankaj Kumar Ojha**
Published by: **BIOTECH BOOKS, NEW DELHI**

11 Application of New Tools in Plant Disease Forecasting

Chandan Maharana and Ch. Yashaswini

Department of Plant Pathology, College of Agriculture, PJTSAU, Rajendranagar, Hyderabad – 500 030, Telengana

Disease Forecasting

Forecasting of plant diseases means predicting for the occurrence of plant disease in a specified area ahead of time, so that suitable control measures can be undertaken in advance to avoid losses. Disease forecasts are predictions of probable outbreaks or increase in intensity of disease. It involves well organized team work and expenditure of time, energy and money. It is used as an aid to the timely application of chemicals. Among the first spray warning services to be established for growers, were the grapevine downy mildew forecasting schemes in France,Germany and Italy in the 1920s.

Information's Needed for Disease Forecasting

Forecasting diseases is a part of applied epidemiology. Hence, knowledge of epidemiology (development of disease under the influence of factors associated with the host, pathogen) is necessary for accurate forecasting. The factors of epidemic and its components should be known in advance before forecasting is done.

The informations required for forecasting are:

1. Host Factors

a. Prevalence of susceptible varieties in the given locality
b. Response of host at different stages of the growth to the activity of pathogen *e.g.* Some diseases are found during seedling stages while others attack grown up plants and

c. Density and distribution of the host in a given locality. Dense populations of susceptible variety invite quick spread of an epidemic. Growing susceptible varieties in scattered locations and that too in a limited area are less prone to epiphytotic.

2. Pathogen Factors

a. Amount of primary (initial) inoculum in the air, soil or planting material
b. Dispersal of inoculum
c. Spore germination
d. Infection
e. Incubation period
f. Sporulation on the infected host
g. Re-dispersal/Dissemination of spores
h. Perennating stages
i. Inoculum potential and density in the seed, soil and air

3. Environmental Factors

a. Temperature
b. Humidity
c. Light intensity
d. Wind velocity

Requirements or Conditions for Disease Forecasting

There are five main requirements which must be satisfied before a useful and successful disease forecast is made.

1. The disease must cause economically significant damage in terms of yield loss or quality.
2. Control measures must be available at an economically acceptable cost.
3. The disease must vary each season in the timing of the first infections and its subsequent rate of progress. If it does not, there is no need for forecasting.
4. The criteria or model used in making a prediction must be based on sound investigational work carried out in the laboratory and in the field and tested over a number of years to establish its accuracy and applicability in all the locations where its use is envisaged.
5. Growers must have sufficient man power and equipments to apply control measures when disease warning is given. Long-term warnings or predictions are more useful than short-term warning or predictions.

The principle behind plant disease forecasting systems is to determine the risk that a disease will occur, or that the intensity of the disease will increase (Campbell and Madden 1990). What defines a successful plant disease forecasting system outlined several attributes, including:

- ✰ Reliability (use of sound biological and environmental data),
- ✰ Simplicity (the simpler the system, the more likely it will be applied and used by producers),
- ✰ Importance (the disease is of economic importance to the crop, but sporadic enough that the need for treatment is not a given),
- ✰ Usefulness (the forecasting model should be applied when the disease and/or pathogen can be detected reliably),
- ✰ Availability (necessary information about the components of the disease triangle should be available),
- ✰ Multipurpose applicability (monitoring and decision-making tools for several diseases and pests should be available), and
- ✰ Cost effectiveness (forecasting system should be cost affordable relative to available disease management tactics).

Methods of Disease Forecasting

Disease forecasting requires field observations on the pathogen characters, collection of weather data, variety of the crop and certain investigations and their correlations. Usually the following methods are employed in disease forecasting. Many plant disease forecasting systems have emphasized forecasts based on the following principles (Agrios 2004; Campbell and Madden 1990):

1. Forecasting Based on Primary Inoculums

Presence of primary inoculum, its density and viability are determined in the air, soil or planting material. Occurrence of viable spores or propagules in the air can be assessed by using different air trapping devices (spore traps). In the case of soil-borne diseases the primary inoculum in the soil can be determined by monoculture method. Presence of loose smut of wheat, ergot of pearlmillet and viral diseases of potato can be detected in the seed lots at random by different seed testing methods. Seed testing methods can be used to determine potential disease incidence and enable decision to be made on the need for chemical seed treatment. The extent of many virus diseases is dependent on the severity of the preceding winter which affects the size of vector population in the growing season. *e.g.*, Sugarbeet yellows virus.

2. Forecasting Based on Weather Conditions

Weather conditions *viz.*,temperature, relative humidity, rainfall, light, wind velocity *etc.*, during the crop season and during the inter crop season are measured. Weather conditions above the crop and at the soil surface are also recorded. *e.g.* potato late blight.

3. Forecasting Based on Correlative Information

Weather data of several years are collected and correlated with the intensity of the diseases. The data are compared and then the forecasting of the disease is done. Forecasting criteria developed from comparisons of disease observation with standard meteorological data have been provided for diseases like *Septoria* leaf blotch of wheat, fire blight of apple and barley powdery mildew.

4. Use of Computer for Disease Forecasting

In some advanced countries forecasting of disease is made by the use of computers. EXPERT systems and Simulation models are included here.This system gives the results quickly. One such computer based programmes in the USA is known as 'Blitecast' for potato late blight.In India its INDOBLIGHTCAST for same pathogen.

5. New Tools: Remote Sensing, GIS and GPS, Thermography

Remote Sensing

Remote sensing and digital image analysis are methods of acquisition and interpretation of measurements of an object without physical contact between the measuring device and the object. The object can be analyzed many times noninvasively and without damage.

The specific properties of the vegetation, healthy or diseased, influence the amount and quality of radiation reflected or emitted from the leaves and canopies; application of this technology to phytopathological research such as phytopathometry is of great interest. Plant pathologists have used remote sensing tools for a number of years and were among the first to use color infrared photography to assess the presence of disease in trees. The application of remote sensing via airborne cameras provided an answer to a question that would not have been possible through ground surveys. In many aspects we have progressed rapidly to our current state of knowledge about the utility of remote sensing. Information from remote sensing can be used as base maps in variable rate applications of fertilizers and pesticides. Information from remotely sensed images allows farmers to treat only affected areas of a field. Problems within a field may be identified remotely before they can be visually identified.

The Electromagnetic Spectrum

The basic principles of remote sensing with satellites and aircraft are similar to visual observations. Energy in the form of light waves travels from the sun to Earth. Light waves travel similarly to waves traveling across a lake. The distance from the peak of one wave to the peak of the next wave is the wavelength. Energy from sunlight is called the electromagnetic spectrum. The wavelengths used in most agricultural remote sensing applications cover only a small region of the electromagnetic spectrum. Wavelengths are measured in micrometers (µm) or nanometers (nm). One um is about.00003937 inch and 1 µm equals 1,000 nm. The visible region of the electromagnetic spectrum is from about 400 nm to about 700 nm. The green color associated with plant vigor has a wavelength that centers near 500 nm.

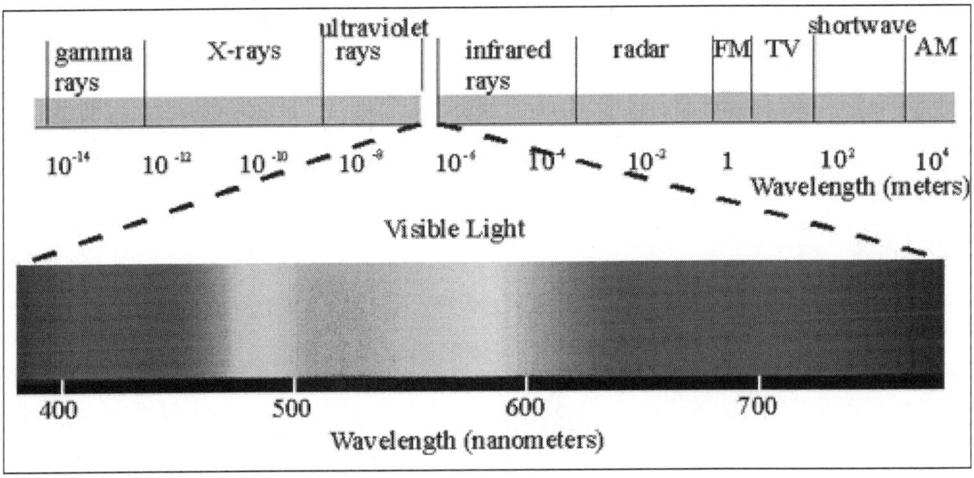

Figure 11.1

Wavelengths longer than those in the visible region and up to about 25 μm are in the infrared region. The infrared region nearest to that of the visible region is the near infrared (NIR) region. Both the visible and infrared regions are used in agricultural remote sensing.

Reflection is that phenomenon in which an impinging beam of radiation of a particular wavelength is reflected back away from the object without any change. This can be contrasted to emittance, which is the emitting of radiant energy at a particular wavelength due to the temperature of an object. Both reflectance and emittance provide information that can be utilized in applying remote sensing to agricultural problems.

The most common source of electromagnetic radiation (ER), which is recorded by satellites, is the sun. When, ER interacts with atmosphere and earth surface (vegetation, soil *etc.*) three things happen

- ✰ Absorption
- ✰ Transmission
- ✰ Reflection

Relationship between these three phenomenon gives a spectral signature which is unique to each plant (Stressed sugarbeets have a higher reflectance value in the visible region of the spectrum from 400-700 nm. This pattern is reversed for stressed sugarbeets in the nonvisible range from about 750-1200 nm. The visible pattern is repeated in the higher reflectance range from about 1300-2400 nm. Interpreting the reflectance values at various wavelengths of energy can be used to assess crop health.)

Depending up on source of illumination remote sensing is of two types *i.e.*

- ✰ Active remote sensing
- ✰ Passive remote sensing

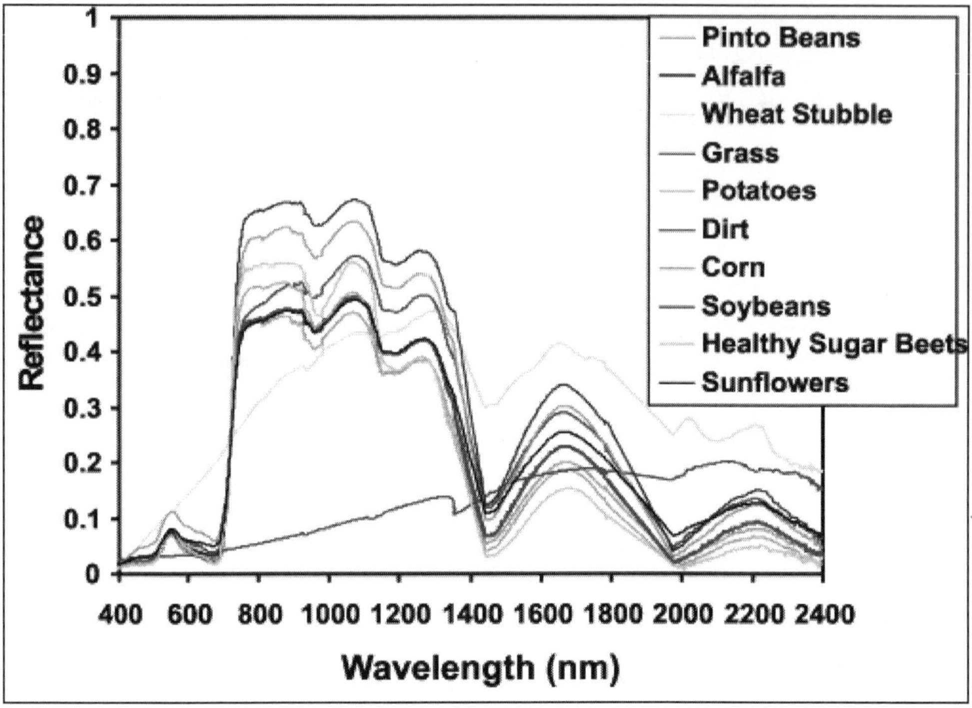

Figure 11.2: Spectral Signatures of Crops and Soil (Kyllo, 2003).

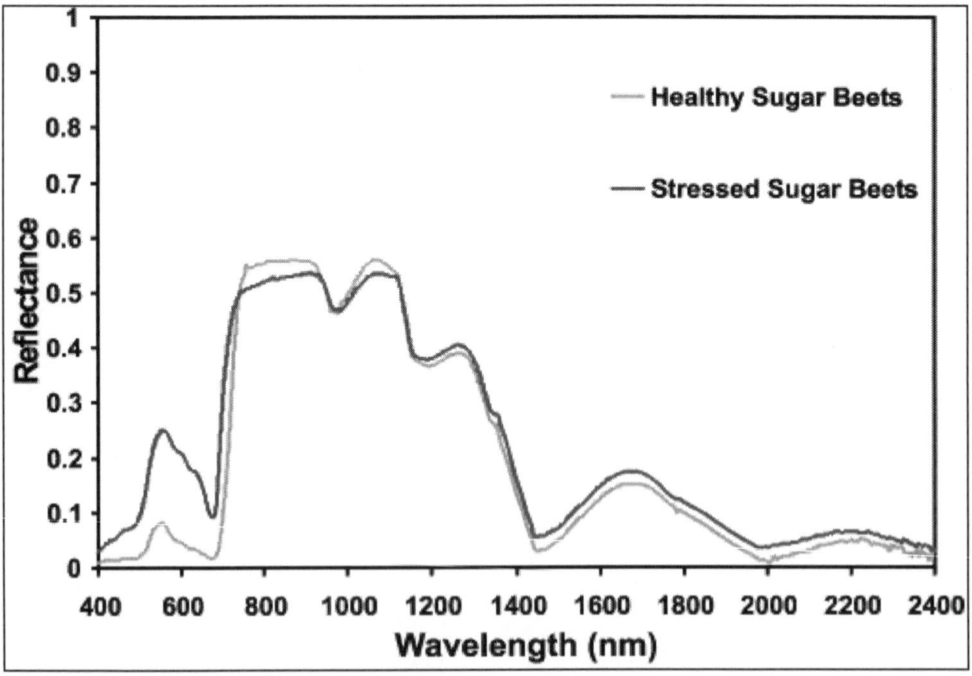

Figure 11.2: Spectral Signatures of Healthy and Stressed Sugarbeets (Kyllo, 2003).

In agriculture we most used passive remote sensing. Optical/Thermal imaging systems can be classified according to the number of spectral bands used:

- ☆ Monospectral or panchromatic (single wavelength band, "black-and-white", grey-scale image) systems
- ☆ Multispectral (several spectral bands) systems
- ☆ Superspectral (tens of spectral bands) systems
- ☆ Hyperspectral (hundreds of spectral bands) systems

Reflection from Leaves

Reflection from individual leaves is not constant across the wavelengths from 400-2500 nm. Leaves have a low reflectance in the visible (400-700nm), a high reflectance in the nearinfrared (700-1200 nm), and a medium reflectance in the middle and far infrared (1200 nm) wavebands. This variation in leaf reflectance has allowed for the differentiation of leaves from soil, which tends to show little variation in reflectance across these wavelengths. Reflectance from leaves is species dependent and sometimes cultivar dependent. The primary variation among species is in the visible reflectance and is due to species or leaf age. Reflectance tends to increase in individual leaves as the leaf matures; however, the changes are wavelength dependent. These changes are due to changes in intracellular water content and chlorophyll content. Lesions and reduction in chlorophyll content created by a disease also cause an increase in reflectance. Water stress by reducing the internal water content increases the reflectance from an individual leaf. Information gathered from individual leaves provides a basic set of information about the mechanism of the changes occurring within a plant; however, to be of practical application it must be extended to a canopy or field level.

Reflection from Canopies and Fields

Composites of leaves or canopies exhibit the same reflective properties of individual leaves; however, there are a series of variables that now must be considered. Leaf orientation, *i.e.*, the arrangement of leaves on the stem and orientation to the sun, provides a source of variation when viewing a canopy compared with an individual leaf. Also, all leaves are not exposed to the same level of incoming radiant energy and often do not reflect back to the sky due to distortions in the leaf surface. Leaf surfaces often act as polarizing filters and reflect back to portions of the sky that are not always detected by viewing the canopy only the vertical direction. However, the information contained in bidirectional and polarized reflectance has yet to be fully exploited in the evaluation of canopy response to stress. Leaf fluorescence is another attribute that has been observed in all plants and can be related to the efficiency of the photosynthetic process. It is possible that leaf fluorescence could be used to assess the impact of diseases on the physiological status of the plant. This technique has only been used on individual leaves; however, it could be extended to canopies through the use of laser-induced fluorescence. This procedure will have to be adapted to plant canopies but may become a powerful and useful research tool. Canopies of plants are grown in fields with varying soil, and soil also has some unique reflective properties. The variation across wavelengths is less for soil than for leaves; however, the reflectance changes in response to modifications in the surface. The addition of organic matter as residue on the surface reduces the reflectance. Soils vary in reflectance due to mineral composition and weathering of the minerals. Instruments available for the measurement of reflected radiation adaptable to remote sensing range from the portable spectro-radiometer, which measures all wavelengths between 400-1100 nm to radiometers with multiple channels set for discrete wave- bands. Instruments with individual channels mimic the wavebands available on the current satellites. A rapidly emerging technology that has yet to be applied is the use of multiple waveband video cameras. This system offers a capability not possible with other radiometers, in that the data are readily available for viewing without intense signal processing and manipulation. Video camera systems may provide a practical tool for disease assessment.

Vegetative Indices

To use the information contained in the reflectance across wavelengths, several vegetative indices have been proposed and evaluated. These indices are based primarily on the ratio or difference between the reflectance in the near-infrared and red wavelengths. The comparison of the reflectance values at different wavelengths, called a vegetative index, is commonly used to determine plant vigor. The most common vegetative index is the normalized difference vegetative index (NDVI). NDVI compares the reflectance values of the red and NIR regions of the electromagnetic spectrum. The NDVI value of each area on an image helps identify areas of varying levels of plant vigor within fields.

Emitted Radiation

All objects that have a temperature emit radiation according to Planck's Law. Soil and plant canopies emit energy, and given the temperatures found on the earth's surface, range in the 1000-1400 nm wavebands. Temperature of a plant canopy can be described by the temperature of individual leaves, the temperature of foliage, or the temperature of the canopy that includes the soil. Leaf temperatures that have been measured relative to the ccurrence of Verticillum wilt or brown rot of soybeans (*Phialo-phora gregata* Gans) have been measured with attached leaf thermocouples. Other measurements of Verticillum wilt have been made with infrared thermometers. Each method has provided a unique relationship of describing the change in leaf temperature relative to the presence of a disease.

GIS (Geographic Information System)

- ☆ A GIS is a computer system capable of capturing, storing, analyzing, and displaying geographically referenced information; that is, data identified according to location.
- ☆ **GEOGRAPHIC** *implies that locations of the data items are known, or can be calculated, in terms of Geographic coordinates* (*Latitude, Longitude*)
- ☆ **INFORMATION** *implies that the data in a GIS are organized to yield useful knowledge, often as coloured maps and images, but also as statistical graphics, tables, and various on-screen responses to interactive queries.*

- **SYSTEM** *implies that a GIS is made up from several inter-related and linked components with different functions. Thus, GIS have functional capabilities for data capture, input, manipulation, transformation, visualization, combinations, query, analysis, modelling and output.*
- Practitioners also define a GIS as including the procedures, operating personnel, and spatial data that go into the system.
- First developed in North America, particularly the U.S. and Canada in the mid-1960s
- Previously been used in natural resources and environmental research.

GPS

It is a device which uses 24 satellites orbiting at 24,200 km above earth to find latitudes, longitudes and altitude of any location on the earth.

Basic Functions of GIS

- Data Acquisition and processing
- Database Management and Retrieval
- Spatial Measurement and Analysis
- Graphic output and Visualization

Other Applications of Remote Sensing and GIS

- Crop pest and disease assessment
- Inventory and mapping of Agricultural lands.
- Assessment of changes in Agricultural land.
- Assessment of irrigated agricultural land.
- Crop inventory, crop hectarage assessment, crop yield estimate.
- Soil categorization and mapping.
- Assessment of degraded lands.
- Land capability/suitability assessment.
- Early detection of crop diseases.
- Food security arrangement.
- Monitoring of crop conditions including weed, herbicide and fertilizer misapplication.
- Estimation of Agrometeorology parameters (rainfall amount and period, temporal).
- Farm planning and plot layout.

There are several types of remote sensing systems used in agriculture but the most common is a passive system that senses the electromagnetic energy reflected from plants. The sun is the most common source of energy for passive systems. Passive system sensors can be mounted on satellites, manned or unmanned aircraft, or directly on farm equipment. There are several factors to consider when choosing a remote sensing system for a particular application, including spatial resolution, spectral resolution, radiometric resolution, and temporal resolution.

FASAL: The Operational Programme for Crop Assessment in India

(*Forecasting Agricultural out put using Space, Agrometeorology and Land based observations,Mahalanobis National Crop Forecast Centre, 2015*)

- ☆ **Crops (8)** Rice (Kharif and Rabi), Jute, Cotton, Sugarcane, Wheat, Rapeseed and Mustard, Soghum(Rabi), Rabi Pulses,
- ☆ **Data** Multidate Microwave (RISAT-1 C band SAR) and Optical (Resourcesat-2 AWiFS and LISS III) Data
- ☆ **Approach** Ground truth using Smartphone, Acreage using satellite data, Yield using weather and remote sensing based models
- ☆ **States Covered** All those states, which together contribute >90 per cent of the Crop's area in the Country
- ☆ **Periodicity** (All forecasts are issued pre-harvest) Paddy, Wheat and Mustard – 3 Forecasts Sugarcane, Cotton – 2 Forecasts Jute, Sorghum and Pulses- 1 Forecast
- ☆ **Software** FASALSoft, developed by ISRO

Conclusion

Remote sensing technique is very useful, less time consuming and cost effective tool for suggesting action plans/management strategies for agricultural sustainability of any region. Remotely sensed images can be used to identify nutrient deficiencies, diseases, water deficiency or surplus, weed infestations, insect damage, hail damage, wind damage, herbicide damage, and plant populations. GIS is considered one of the important tools for decision making in problem solving environment dealing with geo-information.

REFERENCES

Anonymous, 1947. The measurement of potato blight. Transactions of the British Mycological Society 31: 140–141.

Campbell, J. B. 1996. Introduction to remote sensing. London: Taylor and Francis.

Chaerle, L. and Vander S. D. 2000. Imaging techniques and the early detection of plant stress. *Trends in Plant Science* 5: 495-501.

CPRI, 2012. http: //cpri.ernet.in/site/default.aspx (retrived on 29.10.2015)

Gebbers, R. and Adamchuk, V. I. 2010. Precision agriculture and food security. *Science* 327: 828-831.

Hillnhütter, C. and Mahlein, A. K. 2008. Remote sensing to detect plant stress induced by *Heteredora schachtti* and *Rhizoctonia solani* in sugarbeet fields. *Field Crop Research* 122: 70-77.

Jones, H. G. and Schofield, P. 2008. Thermal and other remote sensing of plant stress. *General and Applied Plant Physiology* 34: 19–32.

Laudien, R., Bareth, G. and Doluschitz, R. 2003. Analysis of hyperspectral field data for detection of sugar beet diseases, EFITA 2003 Conference Debrecen, Hungary,pp. 375-381.

Lindenthal, M., Steiner, U., Dehne, H. W., and Oerke, E. C. 2005. Effect of downey mildew development on transpiration of cucumber leaves visualized by digital thermography. *Phytopathology* 95: 233–240.

MacHardy, W.E., 1996. Apple Scab: Biology, Epidemiology and Management. The American Phytopathological Society, St. Paul, MN., ISBN: 0890542066.

MacHardy, W. E., Gadoury D. M. and Rosenberger D. A., 1993. Delaying the onset of fungicides programme for control of apple scab in orchards with low potential ascospore dose of *Venturia inaequalis. Plant Disease* 77: 372-375.

NCFC, 2007. http: //www.ncfc.gov.in/site/223/default.aspx (retrieved on 29.10.2015)

Oerke, E. C. and Dehne, H. W. 2004. Safeguarding production losses in major crops and the role of crop protection. *Crop Protection* 23: 275-285.

Panigrahy, S., Ray, S. S. and Singh, A. 2001. Remot sensing and GIS for potato crop information case studies in India. *South and West Asia Region Internationl Potato Center Newsletter*, 5,4.

Raghava, G. P., Kaundal, R. and Kapoor, A.S. 2006. Machine learning techniques in disease forecasting: a case study on rice blast prediction. *BMC bioinformatics* 7: 485.

Raouse, J.W., Haas, R.H., Schell, J.A. and Deering, D.W.1973. Monitoring vegetation system in great plains with ERTS. Proc. 3rd ERTS-1 Symposium, GSFC, NASA, SP-351, pp. 48-62.

Singh, K.P., Kumar, J., Singh, A., Prasad, R.K., Singh, R.P. and Prasad, D. 2015. Predicting Potential Ascospores Dose of *Venturia inaequalis* (Cks)Wint in farmers apple orchard in central Himalayas of India. *Plant Pathology Journal* 17: 185-195.

Singh, K.P. and Kumar J. 1999. Studies on ascospore maturity of *Venturia inaequalis*, the apple scab pathogen, in Central Himalayas of India. *J. Mycol. Plant Pathol* 29: 408-415.

Stoll, M., Schultz, H. R., Baecker, G. and Berkelmann-Loehnertz, B. 2008. Early pathogen detection under different water status and the assessment of spray application in vineyards through the use of thermal imagery. *Precision Agriculture* 9: 407–417.

Vadivambal, R. and Jayas, D. S. 2011. Applications of thermal imaging in agriculture and food industry—a review. *Food and Bioprocess Technology* 4: 186–199.

Van der Plank, J.C. 1963. Plant Disease: Epidemics and Control. Academic Press, London.

Conservation Agriculture Technologies *Pages* **133–141**
Editor: **Rahul Kumar, Abhinandan Singh and Pankaj Kumar Ojha**
Published by: **BIOTECH BOOKS, NEW DELHI**

12 Rainwater Harvesting (RWH): A Way of Water Management in Rainfed Agriculture

Reena Kumari[1], Babloo Sharma[2] and Pratibha Kumari[3]

[1]Agricultural Engineering Department, N.M. College of Agriculture, Navsari Agricultural University, Navsari – 396 450, Gujarat
[2]Department of Soil and Water Conservation, Faculty of Agriculture, Bidhan Chandra Krishi Viswavidyalaya, Mohanpur, Nadia – 741252, West Bengal
[3]Department of Soil Science and Agricultural Chemistry, Institute of Agricultural Sciences, Banaras Hindu University, Varanasi – 221 005, Uttar Pradesh

Introduction

A large part of the rainfall returns to the atmosphere directly by evaporation from the soil surface and also a part of that infiltrated into the soil to a small depth evaporates into the atmosphere with no benefits. The part that flows as run-off, if not intercepted, goes to slumps, losing its good quality and evaporating; it may even flow into the sea. It was estimated that only about 10 per cent of the annual rainfall on the dry rangelands of West Asia and North Africa (WANA) is beneficially used for supporting vegetation cover, replenishing the groundwater and other purposes. Other factors, such as degraded soils, steep topography, poor vegetative cover and unfavourable climate, besides the poor rain characteristics, aggravate the problem, causing irreversible desertification and detrimental loss of both water and land productivity. Water harvesting (WH) is one option that increases the amount of water per unit cropping area, reduces drought and enables use of run-off beneficially.

There is a need of a more efficient capture and use of the scarce water resources in Rainfed areas. An optimization of the rainfall management, through water harvesting in sustainable and integrated production systems can contribute for improving the small-scale farmers' livelihood by upgrading the rainfed agriculture production. Rainwater harvesting is broadly defined as the collection and concentration of runoff for productive purposes such as crop, fodder, pasture or trees production, livestock and domestic water supply in arid and semi-arid regions. For agriculture purposes, it is defined as a method for inducing, collecting, storing and conserving local surface runoff in arid and semi-arid regions. Water harvesting has been used for many years in different areas all over the world to solve the problem of water scarcity in arid and semi-arid areas.

Rain water harvesting systems have the following characteristics: it is practiced in rainfed regions, where surface runoff often has an intermittent character, it is based on the utilization of runoff and requires a runoff producing area and a runoff receiving area; because of the intermittent nature of runoff events, water storage is an integral part of the system and it can be done directly in the soil profile or in small reservoirs, tanks and aquifers.

The aim of the rainwater harvesting is to mitigate the effects of temporal shortages of rain to cover both household needs as well as for productive use. It has been used to improve access to water and sanitation, improve agricultural production and health care thus contributing to poverty alleviation, reverse environmental degradation through reforestation and improved agriculture practice, aid groundwater recharge, empower women in the management of water and other natural resources and address floods and droughts by storing excess water. In crop production systems, rainwater harvesting is composed of a runoff producing area normally called catchment area and a runoff utilization area usually called cropped basin. The major categories are classified according to the distance between catchment area and cropped basin as follow: *In situ* rainwater harvesting, Internal (Micro) catchment rainwater harvesting and External (Macro) catchment rainwater harvesting.

In situ Rain Water Harvesting

In situ rain water harvesting, also called soil and water conservation, involves the use of methods that increase the amount of water stored in the soil profile by trapping or holding the rain where it falls. In this application, there is no separation between the collection area and the storage area, the water is collected and stored where it is going to be utilized. *In situ* rainwater harvesting involves small movements of rainwater as surface runoff, in order to concentrate the water where it is wanted most. It is basically a prevention of net runoff from a given cropped area by holding rain water and prolonging the time for infiltration. This system works better where the soil water holding capacity is large enough and the rainfall is equal or more than the crop water requirement, but moisture amount in the soil is restricted by the amount of infiltration and or deep percolation.

The first step in any RWH system involves methods to increase the amount of water stored in the soil profile by trapping or holding the rain where it falls. This may involve small movements of rainwater as surface runoff in order to concentrate the water where it is wanted most. *In situ* RWH is sometimes called water conservation and is basically a prevention of net runoff from a given cropped area by holding rain water and prolonging the time for infiltration. This system works better where the soil water holding capacity is large enough and the rainfall is equal or more than the crop water requirement, but moisture amount in the soil is restricted by the amount of infiltration and or deep percolation.

Deep Tillage

Tillage normally assists in increasing the soil moisture holding capacity through increased porosity, increasing the infiltration rates and reducing the surface runoff by providing surface micro-relief or roughness which helps in temporary storage of rain water, thus providing more time for infiltration. Previous research results have shown that the depth of tillage is the most important factor controlling or affecting soil moisture characteristics. Deep tillage helps to increase porosity, reduce surface sealing of the soil and permits roots proliferation to exploit soil water and nutrients at deep horizons.

Conservation Tillage

There are five types of conservation tillage systems: no-tillage, mulch tillage, strip or zonal tillage, ridge till and reduced tillage. The common element in these systems is the presence of crop residue on the soil surface to reduce water and wind erosion. The amount of crop residue may vary widely, but it must be enough to reduce erosion significantly in comparison with tillage systems that bury or remove the residue. The other particularity of conservation tillage systems is that it relies more on herbicides and less on cultivation for weed control, in order to reduce the disturbance or inversion of the soil. This practice also involves ripping the land with tinned implements or sub-soiling the land immediately after crops is harvested, to break the plough pans.

Contour Farming and Ridging

Contour farming involves aligning plants rows and tillage lines at right angles to normal flow of runoff. It creates detention storage in the soil surface horizon and slows down the runoff, thus giving the water time to infiltrate into the soil. This is important where cultivation is done on slopes ranging from 3 per cent to 8. All farm husbandry practices such as tilling and weeding are done along the contours so as to form cross-slope barrier to the flow of water. Where this is not enough it is complemented with ridges which are sometimes tied to create a high degree of surface roughness to enhance the infiltration of water into the soil. The implementation of the system results in less benefit to compacted or poorly permeable soils because these soils become saturated quickly. Also, special skills may be required to construct effective contour lay outs.

Reduced Tillage

The aim of this system is to minimize soil disturbance, while at the same time achieving a viable seedbed for crop growth. As with no-tillage systems, weeds and diseases are usually controlled with herbicides and grazing. Crop residues are usually burnt and or incorporated into the soil. There are three different types of reducing tillage: reduced cultivation, direct drilling and minimum tillage.

No-tillage System

The no-tillage system is specialized type of conservation tillage consisting of a one pass planting and fertilizer operation in which the soil and the surface residues are minimally disturbed. The technique as a system where the soil is not disturbed between harvesting one crop and planting the next. It is a crop production where the soil is not traditional tilled or cultivated though sticks or other planting equipments are used to make the opening for seeds. This is the most effective conservation practice for reducing soil erosion and improving water quality. The crop residue cover and infiltration rates associated with no-tillage maximize the volume reduction of agricultural runoff and contaminants. Weed control is generally achieved with herbicides such as glyphosate or gramoxylin or in some cases with crop rotation. The system eliminates all mechanical seedbed preparation, except for the opening of a narrow (2-3 cm wide) strip or small hole in the ground for seed placement to ensure adequate seed/soil contact.

Micro-catchment RWH

Micro-catchment rain water harvesting system is a method of collecting surface runoff from a small catchment area and storing it in the root zone of an adjacent infiltration area. The system is mainly used for growing medium water demanding crops, such as maize, sorghum, groundnuts and millet. It has also been used to supplement rainfall for native vegetation.

Micro-catchment WH received most research attention to obtain run-off even from light showers with minimum conveyance losses even on flat surfaces. Micro-catchment WH is not economically viable in very dry conditions, with mean annual rainfall less than 100 mm. The problem (for economic viability) is to support a reasonably high per-hectare density of trees. Micro-catchment rainwater harvesting systems has the following characteristics: overland flow harvested from short catchment length, catchment length usually between 1 and 30 cm, runoff stored in soil profile, ratio catchment: cultivated area usually from 1: 1 to 3: 1, normally no provision for overflow and even plant growth. These are the typical examples of this type of system: contour bunds and semi-circular bunds.

The general design principle of Micro-catchment rainwater harvesting systems involves a catchment area, which collects runoff coming from roofs or ground surfaces and a cultivated area, which receives and concentrates runoff from the catchment area for crop water supply. The relationship between the catchment area and the cultivated area, in terms of size, determines by what factor the rainfall will be multiplied. For a more efficient and effective system, it is necessary to calculate the ratio between the two if the data related to the area of concern in terms of rainfall, runoff and crop water requirements is available.

Micro-catchment systems provide many advantages over other irrigation schemes. They are simple and inexpensive to construct and can be built rapidly using local materials and manpower. The runoff water has a low salt content and, because it does not have to be transported or pumped, is relatively inexpensive. The system enhances leaching and often reduces soil salinity. This system is mainly used for growing medium water demanding crops such as maize, sorghum, groundnuts and millet. The major characteristics of the system include:

Pitting System

Pitting system consists of small circular/semicircular pits, with about 30 cm in diameter and 20 cm deep, dug to break the crusted soil surface, to store water and to build up soil fertility. FYM is added in the pits thus permitting the concentration of water and nutrients. Seeds are planted in the middle of the pits. They are used in areas with rainfall of between 350- 600 mm. Organic manure and compost are usually added into the pit to improve fertility. It works by combination of water harvesting and conservation of both moisture and fertility in the pit.

Inter-row Water Harvesting

Inter-row water harvesting is applied either on flat land or on gentle slopes of up to 5 per cent having soil at least 1 m deep. The annual rainfall should not be less than 200 mm/year. On flat terrain (0-1 per cent inclination) bunds are constructed, compacted and, under higher-input conditions, treated with chemicals to increase runoff. The aridity of the location determines the catchment to cropping ratio (CCR), which varies from 1: 1 to 5: 1. On sloping land (1 - 20 per cent inclination) these systems are also called "contour ridges. The ridges of about 0.40 m height are built 2 to 20 m apart, depending on slope, soil surface treatment, general CCR and type of crop to be grown. The catchment area should be weeded and compacted; the crops are either grown in the furrow, along the upper side of the bund or on top of the bund. On sloping land, this system is recommended only for areas with a known regular rainfall pattern; very high rainfall intensities may cause breakages of the bunds. Crops cultivated in row water harvesting systems are maize, beans, millet, rice or (in the USA) grapes and olives.

Contour Bunds

This system consists of small trash, earth or stone embankments, constructed along the contour lines. The embankments trap the water flow behind the bunds allowing deeper infiltration into the soil. The height of the bund determines the net storage of the structure. The water is stored in the soil profile and above ground to the elevation of the bund or overflow structure. This is a versatile system for crop production in a variety of situations. They can be easily constructed but they are limited to availability of power (for earth moving), stones and trash. They are useful where ground slope is not more than 5 per cent, soil depth is at least 1m and rainfall intensity is less than 20 mm h^{-1} duration storms with P = 20 per cent.

Semi-circular Bunds

These are constructed in series in staggered formation runoff water is collected within the hoop from the area above it and impounded by the depth decided by the height of the bund and the position of the tips. Excess water is discharged around the tips and is intercepted by the second row and so on. Normally the semi-circles are of 4-12 m radius with height of 30 cm, base width of 80 cm, side slopes 1: 1.5 and crest width of 20 cm. The percentage of enclosed area which is cultivated depends on the rainfall regime of the area. Basic requirements of the semi-circular bunds are: a). ground slope must be less than 3 per cent, b). Soil depth, at least 1 m, c). Average annual rainfall of at least 100 mm, d). CBAR of at least 3: 1 and e). Rainfall intensities of I60 equal to 50 mm h^{-1} for rainfall of P = 20 per cent.

Macro Catchment RWH

This is a system that involves the collection of runoff from large areas which are at an appreciable distance from where it is being used. This is sometimes used with intermediate storage of water outside the contour bunds for later use as supplementary irrigation. This system involves harvesting of water from catchments of areas ranging from 0.1 ha to thousands of hectares either located near the cropped basin or long distances away. The catchment areas usually have slopes ranging from 5-50 per cent, while the harvested water is used on cropped areas which are either terraced or on flat lands. When the catchment is large and located at a significant distance from the cropped area the runoff water is conveyed through structures of diversion and distribution networks. The most important systems include the following: Hillside sheet/rill runoff utilization and Floodwater harvesting within the stream bed.

Impact of Harvested Water on Crop Yield

National Agricultural Technology Project (NATP) identified five major production systems *viz.*, coarse cereals based, groundnut based, rainfed rice based, cotton based and soybean based production systems based on soil, climate and cropping pattern for identifying technologies to enhance productivity. Improvement of agricultural productivity in the dry areas goes through the development of land management practices. Water harvesting is thus based on the dryland management principle that aims to deprive part of the land of its low and unproductive share of rain in order to add it to another part of the land and obtain economic yields. Successful implementation of WH practices requires significant knowledge input from hydrology, agronomy and sociology. Data on rainfall, soil and relief, as well as information on the cropping systems and the local socio-economic conditions, are all needed. Identification of areas suitable for WH practices is crucial for successful development of WH. Because WH deals with large areas, the cost of surveys and analysis could be prohibitive impact of irrigation along with other management practices is considerably higher. Non-availability of irrigation at critical stages of plant growth and low efficiency of the canal irrigation systems may be the possible reasons for lower district level yields even under irrigated conditions.

There is a scope for large yield improvements under rainfed conditions. The potential yields that can be attained and the actual yields for some of the dryland crops are presented in Table 12.1. With advanced management and high input levels, yields can be improved two to five times the current levels. Specialized dryland management practices such as water harvesting and reduction of soil moisture loss can increase yields by an additional 5–15 per cent on average across the SAT regions and reduce the variability in yields from year to year, producing a more reliable yield. Besides, the potential productivity varies with the rainfall and the crop growing season of a region. In the case of maize, the potential productivity for the high rainfall regions is 8020 kg/ha, whereas the current national average is about 2062 kg/ha, indicating an unbridged yield gap of 5952 kg/ha. There are many districts in India where the actual yields are lower than the national average and enormous potential for improvement exists. In the case of rainfed rice, which contributes 55 per cent of the total area under rice, the average yield is about 1 t, and the extent of adoption of high-yielding varieties is to the extent of 32 per cent and the achievable yield is to the tune of 2500 kg/ha.

Table 12.1: Potential and Attainable Yields (kg/ha) of some of Rainfed Crops*

Crop	*Potential Yields Attainable*	*Actual Yields (TE 2007–08)*	*Difference*	*Quotient*
Sorghum	4560	902	3658	5.0
Maize	3870	2062	1808	1.9
Pearl millet	2870	906	1964	3.2
Groundnut	2590	1171	1419	2.2
Soybean	2850	1089	1761	2.6

* *Source*: Venkateswarlu and Prasad (2012).

Supplemental Irrigation with Harvested Water: Crop Yield Impact

Supplemental irrigation has been described as a technique used on crops that can be grown using rainfall alone, which applies a limited amount of water during times of low rainfall to ensure that enough water is received to support crop growth and stabilize yields. The goal of supplemental irrigation is to provide enough water during critical growth stages to produce optimal yield per unit of water, not to provide stress-free conditions throughout the growing season with the aim of producing maximum yield. Tables 12.2 and 12.3 presents critical growth stages of important crops and effect of critical irrigation on yield of rainfed crops at different locations in India respectively.

In rainfed areas, WH systems can provide additional water to supplement rainfall to increase and stabilize production. Furthermore, it can alleviate the risk associated with the unpredictability of rainfall in these areas. For this case, the WH system is usually equipped with a facility (above- or underground type) to store the harvested water for later use in supplemental irrigation during drought periods. The water related challenge in rainfed agriculture is to manage the high spatial and

Table 12.2: Critical Growth Stages of Important Crops

Crop	*Important Moisture Sensitive SAtages*
Rice	Panicle Initiation, Flowering
Wheat	Crown root initiation, Jointing, Milking
Maize	Silking, Tasseling
Pearl millet	Flowering, Panicle Initiation
Finger millet	Panicle Initiation, Flowering
Groundnut	Rapid flowering, Pegging, Early pod formation
Redgram Blackgram Greegram	Flowering and Pod formation
Sunflower	Two weeks before flowering and Two weeks after flowering
Soyabean	Blooming and Seed formation
Cotton	Flowering and Boll development
Sesamum	Blooming stage to maturity

Table 12.3: Effect of Critical Irrigation on Yield Rainfed Crops at different Locations in India**

Location	*Crop*	*Yield (t/ ha) Without Irrigation*	*Yield (t/ha) with Critical Irrigation*	*Per cent Increase with Critical Irrigation (Ratio of irrigated versus rainfed yield)*
Ludhiana (4)*	Wheat	1.92	4.11	114.06 (2.14)
Agra(2)	Wheat	2.19	2.74	25.14 (1.25)
Dehradun(4)	Wheat	2.14	3.55	65.89 (1.66)
Rewa(4)	Wheat	0.57	1.88	229.82 (3.30)
Varanasi(2)	Barley	2.60	3.36	29.23 (1.29)
Bijapur (5)	Sorghum	1.65	2.36	43.03 (1.43)
Bellary (4)	Sorghum	0.43	1.37	218.60 (3.19)
Sholapur (5)	Sorghum	0.98	1.82	85.71 (1.86)
Rewa (4)	Upland rice	1.62	2.78	71.60 (1.72)

* Figures in parenthesis indicate average number of seasons.

** *Source*: Reports of All India Coordinated Research Project on Dryland Agriculture, Hyderabad.

temporal availability of rainfall, which increases with lower average totals. There is generally enough rainfall in most of the moist semi-arid and dry sub-humid zones to meet the crop water needs. India should not have to suffer from droughts, if local water balances were managed better. Even during drought years watershed development efforts of improving rainfall management has benefited Indian farmers. National network on Model Watersheds had convincingly established that runoff to a limited extent can be harvested and recycled to stabilize crop production across different climatic zones and production systems. This component of water harvesting has become the backbone in furthering the watershed programs in rainfed areas in most states of India. The available runoff can be harvested and utilized broadly for two purposes- to provide supplemental irrigation to the standing *kharif* crop to offset mid-season dry spells/terminal drought (flowering- grain filling stage) or facilitate sowing of the next *rabi* crop.

REFERENCE

Venkateswarlu B. and Prasad J. V. N. S. (2012). Carrying capacity of Indian agriculture: issues related to rainfed agriculture. *Current Science* 102(6): 882-888.

Conservation Agriculture Technologies *Pages* **143–159**
Editor: **Rahul Kumar, Abhinandan Singh and Pankaj Kumar Ojha**
Published by: **BIOTECH BOOKS, NEW DELHI**

13 Conservation Agriculture for Enhancing Resource Use Efficiency and Sustainability

Adyant Kumar, Anannya Ghosh, Dibyendu Mondal, Ratikanta Ghosh and Pintoo Bandopadhyay

Department of Agronomy, Bidhan Chandra Krishi Viswavidyalaya, Mohanpur – 741252, West Bengal

India is endowed with a rich and vast diversity of natural resources, particularly soil, water, weather, multipurpose trees and bio-diversity. To realize the potential of production system on a sustained basis, efficient management of the natural resources is very crucial. With the advent of high-yielding crop varieties and intensive cultivation, the food grain production has increased from 51 m t in 1950-51 to 264.77 m t during 2013-14. But without developing proper region based resources and the basic needs of intensive agriculture, the different revolutions were made; as a result, India never achieved sustainable productivity excepting in wheat during 70s' and hence, today we are facing food, nutrition, ecological and health insecurity. To meet the varied growing demands for food, fuel, fiber, feed, fertilizer and products in the recent year, the natural resources are however, put under intense strain resulting in fast degradation and lowering of their production efficiency. Besides the increasing industrialization and urbanization and global warming all are dwelling tremendous strain on the shrinking resources. There is 150 m ha degraded land constituting 45.5 per cent of total geographical area. The area sufferings due to water and wind erosion are 109.7 and 11.7 m ha respectively. Conservation agriculture (CA) is an approach for designing and management of sustainable and resource-conserving agricultural systems. It seeks to conserve, improve and make more efficient use of natural resources through integrated management of soil, water, crops and other biological resources in combination with selected external inputs. Elements of Conservation Agriculture, inter-alia, include improved on-farm water management (watershed, global water technology), reduced tillage, organic soil cover, direct seeding using mulching and appropriate crop rotations to avoid pest problems. Over the past few years, adoption of zero tillage and CA has expanded to cover about 1.5 million hectares. In India, CA adoption is still in the initial phases. The direct benefits to farmers of CA are two folds viz. reduced cost of cultivation through savings in seed, water, labour, time and farm power and improved use efficiency resulting in reduced use of resources, the system intensification. In India, efforts to promote CA technologies have largely been focused in the Indo-Gangetic plains covering the states of Punjab, Haryana, Uttar Pradesh, Bihar and West Bengal. Some research works on CA under IRCTs at SAUs, IRRI, particularly reduced tillage on jute, potato, mustard, wheat after rice, SRI, aerobic rice technology, precision land leveling, controlled traffic farming, bed and furrow planting, etc. also showed CA adoption to farmers are gradually increasing though it needs more awareness and demonstration to show the all-round benefits of CA for further progress.

Keywords: *RCTs, Land degradation, Conservation tillage, Sustainability, Food security.*

Introduction

India is endowed with a rich and vast diversity of natural resources, particularly soil, water, weather, multipurpose trees and bio-diversity. To realize the potential of production system on a sustained basis, efficient management of the natural resources is very crucial. With the advent of high-yielding crop varieties and intensive cultivation, the food grain production has increased from 51 m t in 1950-51 to a record figure of 264.77 m t during 2013-14 (Anonymous, 2014).This impressive achievement has pulled the country in to self-sufficiency for food demand. With adoption of intensive agriculture to meet the varied growing demands for food, fuel, fiber, feed, fertilizer and products in the recent year, the natural resources are however, put under intense strain resulting in fast degradation and lowering of their production efficiency. The demographic pressure is rapidly mounting on the natural resources.

The present population of 1 billion, accounting for about 18 per cent of Worlds population supported by only on 2.4 per cent land area, is estimated to become 1.4 billion by 2025 and 1.7 billion by 2050 AD, needing annually about 380 m t and 480 m t food grain respectively (Yadav and Singh, 2000). This scenario along with the increasing industrialization and urbanization will place tremendous strain on the shrinking resources. Unless corrective measures are taken, there will be irreversible damage to the environment and the very resource base. Among the natural resources, soil is a finite, non-elastic and non-renewable asset. The dwindling per caput availability of land that decreased from 0.5 ha in 1950-1951 to 0.15 ha in 1999-2000 and is likely to reduce further to 0.08 ha in 2020 AD. Approximately 80 per cent of the total farm holdings fall in the category of small and marginal with an average holding of <1 ha. Research findings have only shown 3-4 fold productivity potential of land even with the currently available technologies. Thus land of properly managed shall not be a serious constraint to feed the population. The country faces a shortage of water, a key input for agricultural production, despite annual average precipitation of 400 m ha-m, due to erratic and uncertain rainfall in terms of both space and time as also of faculty water management. A substantial portion of groundwater in arid and semi-arid regions and in some humid coastal areas is of poor quality. The growing quantitative demand for freshwater from other sectors *viz.*, industry, power and household are further curtailing agriculture share.

Despite experiencing water scarcity, hazards of its misuse have not received due attention in practice. In areas of freshwater aquifers, there has been excessive withdrawl of groundwater. Judicious integration of all the water resources at the farm, system and basin levels and also appropriate methods for safe use of poor quality water including sewage water will be essential for sound irrigation management.

Land degradation is a major threat to our food and environmental security. There is 150 m ha degraded land constituting 45.5 per cent of total geographical area. The area sufferings due to water and wind erosion are 109.7 and 11.7 m ha respectively. The area under water logging, salinization/alkalization and other problems are 9.0, 9.2 and 10.3 m ha respectively. The widespread erosion in the hilly catchments area is resulting in excessive siltation of multipurpose reservoirs and other water-bodies of the country at rates higher than their designed capacity.

The development of major and medium irrigation works has also caused degradation of large area with the occurrence of water-logging and secondary salinity. About 4.5 m ha area of commands in irrigation is affected by water logging and salinity. Presently, groundwater is used to irrigate almost about 5.5 m tube wells and more than 10 m dug wells. Unfortunately, these advantages have lead to over exploitation of groundwater forcing fall in the groundwater table in most of the areas.

A major factor responsible for the degradation of the natural resources is soil erosion. The accelerated soil erosion has irreversible destroyed some 430 m ha of land area covering 30 per cent of the present cultivated area in different countries of world. In general soil erosion is more severe in mountainous than undulating areas. The rate of natural erosion for the world is of the order of 1.5 to 7.0 t ha^{-1} $year^{-1}$ for the mountainous region and 0.1-7 t ha^{-1} $year^{-1}$ for the undulation plains. If the global warming trends caused by increases in atmospheric CO_2, expected to reach 600 ppm by 2070, continues, global erosion rate may increase considerably. Erosion by water is most serious degradation problem in the Indian context. It has been estimated that soil erosion was taking place at an average rate of 16.35 t/ha/year totalling 5,334 m t/year, Nearly 29 per cent of the total eroded soil was parentally loss to the sea an nearly 10 per cent was deposited in reservoirs, resulting in the reduction of their storage capacity by 1-2 per cent annually. The remaining 61 per cent of the eroded soil was transferred from one place to another. The annual water erosion rate values ranged from <5 t/ha/year to more than 80 t/ha/year.

Conservation agriculture (CA) is concept for resource saving agricultural crop production that strives to achieve acceptable profits together with high and sustainable production levels. While concurrently conserving the environment, CA is based on enhancing natural biological processes above and below ground. Intervention *viz.*, mechanical soil tillage are reduced to an absolute minimum and the use of external inputs such as agrochemicals and nutrients of mineral or organic origin are applied at an optimum level and in a way and quantity that does not interfere to disrupt the biological processes.

What is Conservation Agriculture?

Conservation Agriculture is an approach for designing and management of sustainable and resource-conserving agricultural systems. It seeks to conserve, improve and make more efficient use of natural resources through integrated management of soil, water, crops and other biological resources in combination with selected external inputs. Such a technological package represents a resource saving and efficient agriculture that contributes to environmental conservation and at the same time enhances production on a sustainable basis. Elements of Conservation Agriculture, *inter-alia,* include improved on-farm water management, minimum tillage, organic soil cover, direct seeding through the crop residue and appropriate crop rotations to avoid disease and pest problems. Burning of crop residues, a common practice in many areas (*e.g.* rice-wheat cropping system) further causes pollution, green house gases (GHGs) emission and loss of valuable plant nutrients. When crop residues are retained on the soil surface in combination with no tillage, it initiates processes that lead to improved soil quality and overall resource enhancement.CA has emerged as a new paradigm to achieve goals of sustainable agricultural production. It is a major step towards transition to sustainable agriculture. Therefore, benefits of CA are several folds. Direct benefits to farmers include

- ✰ Reduced cost of cultivation through savings in labour,
- ✰ Time and farm power, and
- ✰ Improved use efficiency resulting in reduced use of inputs.

More importantly, CA practices reduce resource degradation. CA leads to sustainable improvements in efficient use of water and nutrients by improving nutrient balance and availability, infiltration and retention by the soil, reducing water loss due to evaporation and improving the quality and availability of ground and surface water.

It has the potential to emerge as an effective strategy to the increasing concerns of serious and widespread natural resources' degradation and environmental pollution, which accompanied the adoption and promotion of green revolution technologies. Since the early seventies, CA has emerged as a way of transition to the sustainability of intensive production systems. Unlike the rest of world, in India and in South Asian countries, spread of CA technologies is taking place in the irrigated Indo-Gangetic Plains where rice-wheat cropping dominates.

Principles of Conservation Agriculture

Conservation agriculture practices perused in many parts of the world are built on ecological principles making land use more sustainable (Lal, 2013). Adoption of CA for enhancing Resource use efficiency (RUE) and crop productivity is the need of the hour as a powerful tool for management of natural resources and to achieve sustainability in agriculture. Conservation agriculture basically relies on 3 principles, which are linked and must be considered together for appropriate design, planning and implementation processes. These are:

- ✰ **Minimal Mechanical Soil Disturbance:** The soil biological activity produces very stable soil aggregates as well as various sizes of pores, allowing air and water infiltration. This process can be called "biological tillage" and it is not compatible with mechanical tillage. With mechanical soil disturbance, the biological soil structuring processes will disappear.

Minimum soil disturbance provides/maintains optimum proportions of respiration gases in the rooting-zone, moderate organic matter oxidation, porosity for water movement, retention and release and limits the re-exposure of weed seeds and their germination (Kassam and Friedrich, 2009).

- ☆ **Permanent Organic Soil Cover:** A permanent soil cover is important to protect the soil against the deleterious effects of exposure to rain and sun; to provide the micro and macro organisms in the soil with a constant supply of "food"; and alter the microclimate in the soil for optimal growth and development of soil organisms, including plant roots. In turn it improves soil aggregation, soil biological activity and soil biodiversity and carbon sequestration (Ghosh *et al.*, 2010).
- ☆ **Diversified Crop Rotations:** The rotation of crops is not only necessary to offer a diverse "diet" to the soil micro organisms, but also for exploring different soil layers for nutrients that have been leached to deeper layers that can be "recycled" by the crops in rotation. Furthermore, a diversity of crops in rotation leads to a diverse soil flora and fauna. Cropping sequence and rotations involving legumes helps in minimal rates of build-up of population of pest species, through life cycle disruption, biological nitrogen fixation, control of off-site pollution and enhancing biodiversity (Kassam and Friedrich, 2009).

Conservation Agriculture: Indian Scenario

In India, efforts to adapt and promote resource conservation technologies have been underway for nearly a decade but it is only in the past 4 to 5 years that the technologies are finding rapid acceptance by the farmers. Efforts to develop and spread conservation agriculture have been made through the combined efforts of several State Agricultural Universities, ICAR institutes and the CG promoted, Rice-Wheat Consortium for the Indo-Gangetic Plains. Unlike, in the rest of the world, in India spread of technologies is taking place in the irrigated regions in the Indo-Gangetic plains where rice-wheat cropping system dominates.CA systems have not been tried or promoted in other major agro-ecoregions like rainfed semi-arid tropics, the arid regions or the mountain agro-ecosystems.

Rainfed semi-arid and arid regions are characterized by variable and unpredictable rainfall, structurally unstable soils and low overall productivity. Results of most research station studies show that zero/reduced tillage system without crop residues left on the soil surface have no particular advantage because much of the rainfall is lost as runoff due to rapid surface sealing nature of soils. It would therefore appear that no tillage alone in the absence of soil cover is unlikely to become a favoured practice. However, overall productivity and residue availability being low and demand of limited residues for livestock feed being high also poses a major limitation for residue use as soil cover in the arid and semi-arid regions. In the semi-arid regions there is wide variability in rainfall and its distribution and nature of soils. It would appear that there is need to identify situations where availability of even moderate amount of residues can be combined with reduced tillage to enhance soil quality and efficient use of rainwater. There appears no doubt that managing zero-tillage system requires a higher level of management vis-a-vis conventional crop production systems. Also there exists sufficient knowledge to show that benefits of CA mainly consist of reversing the process of degradation and that its advantage in terms of crop productivity may accrue only gradually.

In India efforts to promote CA technologies have largely been concentrating in the Indo-Gangetic plains covering the states of Punjab, Haryana, Uttar Pradesh, Bihar and West Bengal. Over the decades of seventies and eighties rice followed by wheat has emerged the major cropping system extending to over 10 million ha, although a number of other crops are grown in spatial and temporal variations throughout the plains according to agro-climatic variations. Concerns of stagnating productivity, increasing production costs, declining resource quality, declining water table and increasing environmental problems are the major forcing factors to look for alternative technologies particularly in the north-west region encompassing the states of Punjab, Haryana and western UP. In the eastern region covering eastern UP, Bihar and West Bengal developing and promoting strategies to overcome the constraints responsible for continued low cropping system productivity have been the primary concern. It is in this context that the efforts aimed at developing and promoting CA technologies over the past decade or so must be viewed.

The primary focus of developing and promoting CA practices has been development and adoption of zero-till seed cum fertilizer drill for sowing wheat crop in the rice wheat system. Other interventions being tested and promoted include raised bed planting system, laser equipment aided land levelling, residue management alternatives, alternative to rice-wheat cropping system in relation to CA technologies, *etc.* The area planted with wheat adopting zero-till drill has been rapidly increasing over the past five to six years. A study from Kurukshetra district in Haryana showed that the number of zero-till cum fertilizer drills increased from 5 in 1997 to 1425 in 2003 and the corresponding area sown with wheat increased from 50 ha to 40,000 ha. The study also brought out that adoption of this technology was not limited to farmers with a large holding but the small holding farmers were adopting the technology through custom hiring (Rs 750 to 900 per ha for sowing wheat) while the farmers with bigger holding are purchasing the seed drill. In a sample study amongst farmers adopting zero-tillage it was observed that the area brought under zero-tillage *vis-a-vis* total area under wheat crop varied with an overall average of 45 per cent under zero-till system.

Another study summarized results of more than six hundred demonstrations over an area of 440 ha from 1998 to 2000. These and several other studies concluded that wheat yields in zero-till fields were invariably higher by 3 to 5 per cent compared to planting in conventionally tilled fields. Increased yields were largely a result of timely planting of wheat when zero-till was used. Reduced cultivation costs to the tune of Rs 1500 to 2250 per ha resulting from reduced costs of diesel, labour and use of herbicides were other advantages cited by most studies presented. Reduced incidence of weeds, particularly *Phalaris minor* and savings in water use were other benefits frequently cited by participants. In addition to zero-till planting of wheat, raised bed planting and laser land levelling are other technologies being increasingly adopted by farmers of the region. Farmers who have used laser land leveller reported positive effects in improving crop establishment, uniformity of crop maturity, increase in area available for cultivation (2 to 5 per cent), improved efficiency of water application and increased water productivity resulting in large savings in irrigation water (up to 35 per cent) and in improved use efficiency of applied nutrients. With increased availability of laser land levelling equipment, precision land levelling could become an important basal step for adopting CA systems.

Status of Conservation Agriculture in India and Abroad

Globally, CA is being practiced on about 125 M ha (Table 13.1). The major CA practicing countries are USA (26.5 M ha), Brazil (25.5 M ha), Argentina (25.5 M ha), Canada (13.5 M ha) and Australia (17.0 M ha). In India, CA adoption is still in the initial phases. Over the past few years, adoption of zero tillage and CA has expanded to cover about 1.5 million hectares (Jat *et al.*, 2012). The major CA based technologies being adopted is zero-till (ZT) wheat in the rice-wheat (RW) system of the Indo-Gangetic plains (IGP). In other crops and cropping systems, the conventional agriculture based crop management systems are gradually undergoing a paradigm shift from intensive tillage to reduced/zero-tillage operations. In addition to ZT, other concept of CA need to be infused in the system to further enhance and sustain the productivity as well as total new sources of growth in agricultural productivity. The CA adoption also offers avenues for much needed diversification through crop intensification, relay cropping of sugarcane, pulses, vegetables *etc.* as intercrop with wheat and maize and to intensify and diversify the RW system. The CA based resource conservation technologies also help in integrating crop, livestock, land and water management research in both low-and high-potential environments.

Table 13.1: Global Adoption of Conservation Agriculture Systems

Continent	*Area (m ha)*	*Percentage*
United State of America	26.5	21.2
Brazil	25.5	20.4
Argentina	25.5	20.4
Australia and New Zealand	17.0	13.6
Canada	13.5	10.8
Russia Federation	4.5	3.6
China	3.1	2.5
Paraguay	2.4	1.9
Kazakhstan	1.6	1.3
Others	5.3	4.2
World's Total	124.8	100

Source: FAO, 2012.

In India, efforts to adopt and promote conservation agriculture technologies have been underway for nearly a decade but it is only in the last 8 – 10 years that the technologies are finding rapid acceptance by farmers. Efforts to develop and spread conservation agriculture have been made through the combined efforts of several State Agricultural Universities, ICAR institutes and the Rice-Wheat Consortium for the Indo-Gangetic Plains. The spread of technologies is taking place in India in the irrigated regions in the Indo-Gangetic plains where rice-wheat cropping systems dominate. Conservation agriculture systems have not been tried or promoted in other major agro-eco regions like rainfed semi-arid tropics and the arid regions of the mountain agro-ecosystems. Spread of these technologies is taking place in the irrigated regions of the Indo-Gangetic plains where the rice-wheat cropping system dominates. The focus of developing and promoting conservation technologies has been on zero-till seed-cum fertilizer drill for sowing of wheat in rice-wheat system. Other interventions include raised-bed planting systems, laser equipment aided land leveling, residue management practices, alternatives to the rice-wheat system *etc.* It has been reported that the area planted with wheat adopting the zero-till drill has been increasing rapidly (Malik *et al.*, 2005b), and presently 25 per cent – 30 per cent of wheat is zero-tilled in rice-wheat growing areas of the Indo-Gangetic plains of India. In addition, raised-bed planting and laser land leveling are also being increasingly adopted by the farmers of the north-western region.

Impact of Conservation Agriculture Practices

The rapid adoption and spread of CA technologies particularly zero-tillage for wheat is attributed to multiplicity of benefits. These include:

- ☆ **Reduction in Cost of Production:** This is a key factor contributing to rapid adoption of zero-till technology. Most studies show that the cost of wheat production is reduced by Rs.1500 to 2000 per hectare. Cost reduction is attributed to savings on account of diesel, labour and input costs, particularly herbicides.
- ☆ **Reduced Incidence of Weeds:** Most studies tend to indicate reduced incidence of *Phalaris minor*, a major weed in wheat, when zero-tillage is adopted resulting in reduced use of herbicides.
- ☆ **Saving in Water and Nutrients:** Limited experimental results and farmers experience indicate that considerable saving in water (up to 20-30 per cent) and nutrients are achieved with zero-till planting and particularly in laser levelled and bed planted crop.
- ☆ **Increased Yields:** In properly managed zero-till planted wheat yields were invariably higher by 4 to 6 per cent compared to traditionally prepared fields for comparable planting date.
- ☆ **Environmental Benefits:** CA involving zero-till and surface managed crop residue systems is an excellent opportunity to eliminate burning of crop residues which contribute to large amount of green house gases like CO_2, CO, NO_2, SO_2 and large amount of particulate matter. Burning of crop residues, also contributes to considerable loss of plant nutrients, which could be recycled when properly managed. Large scale burning of crop residues is also a serious health hazard.

- **Crop Diversification Opportunities:** Adopting CA system (includes planting on raised beds) offers opportunities for crop diversification. Cropping sequences/rotations and agro-forestry systems when adopted in appropriate spatial and temporal patterns can further enhance natural ecological processes which contribute to system resilience and reduced vulnerability to yield reducing disease/pest problems. Limited studies indicate that a variety of crops like mustard, chickpea, pigeonpea, sugarcane, *etc.*, could be well adapted to the new systems with advantage.
- **Resource Improvement:** No tillage when combined with surface managed crop residues sets in the processes whereby slow decomposition of residues results in soil structural improvement and increased recycling and availability of plant nutrients. Surface residues acting as mulch, moderate soil temperatures, reduce evaporation, improve biological activity and provide more favourable environment for root growth, the benefits which are traditionally sought from tillage operations.

Resource Conserving Technologies

The term conservation agriculture (CA) and resource-conserving technologies (RCTs) were used as if their meanings were similar. Henceforth, resource conserving technologies will refer to those practices that enhance resource- or input-use efficiency. This covers a lot of ground. New varieties that use nitrogen more efficiently may be considered RCTs. Zero or reduced tillage practices that save fuel and improve plot-level water productivity may also be considered RCTs, as may land leveling practices that help save water. There are many, many more. In contrast, conservation agriculture practices will only refer to the RCTs with the following characteristics:

- Soil cover, particularly through the retention of crop residues on the soil surface;
- Sensible, profitable rotations; and
- A minimum level of soil movement, *e.g.*, reduced or zero tillage.

The distinction is important because some RCTs, while attractive in the near-term, may be unsustainable in the longer-term. An example of this is the use of zero tillage without residue retention and without suitable rotations which, under some circumstances, can be more harmful to agro-ecosystem productivity and resource quality than a continuation of conventional practices (Sayre, 2000).

The principles listed above are defined as common to CA systems. However, the specific components of a CA system (establishment methods, farm implement selection, crops in the rotation, soil fertility management, crop residue and mulch management, germplasm selection, *etc.*) tend to be different across environments. Local investments in adaptive research typically are needed to tailor CA principles to local conditions. This process of tailoring is most efficient when an innovation system emerges and begins to acquire a self-sustaining dynamic. When this happens,

"......technology development and adoption [become a] social phenomenon in which agents interact in several ways, creating multiple information flows in many directions. These agents (*e.g.*, public research and extension systems, innovative farmers, commercial firms, foreign research institutions) form networks that co-evolve with the technologies that they create." (Ekboir, 2002, p. 2).

Innovations of Resource Conservation Technologies (RCTs)

The resource conservation technologies that proved successful in India include watercourse improvement, laser land leveling, zero tillage technology, bed and furrow, irrigation system/bed planting, *etc.* These techniques have been promoted on pilot basis for efficient utilization of water and other inputs for crop production.

Laser Land Leveling

For surface irrigated areas it is essential to have a properly levelled surface with the appropriate inclination for the irrigation method adopted. Traditional farmers' methods for levelling by eyesight are not sufficiently accurate (particularly on larger plots), resulting in extended irrigation times, unnecessary water consumption and inefficient water use. The use of laser-guided equipment for the levelling of surface-irrigated fields has become economically feasible and accessible – through hiring services – even to lower-income farmers. Laser levelling reduces the unevenness of the field to about ±2 cm, resulting in better water application and distribution efficiency, improved water productivity, increased fertilizer efficiency and reduced weed pressure. Savings of up to 50 per cent in wheat and 68 per cent in rice have been reported (Jat *et al.*, 2006).

Precision land levelling is a resource conservation technology, initially, bucket type soil scrapers were used for precision land levelling, which have now been replaced by laser beam guided automatic scrapers for more precision of land levelling work. Impact assessment studies reveal significant benefits of precision land levelling (Table 13.2). Keeping in view benefits, it warrants that laser land levelling services are strengthened so that common farmers may harvest the fruits of this modern technique.

Table 13.2: Benefits of Laser Land Levelling

Sl.No.	*Particulars*	*Percentage (per cent)*
1.	Curtailment in irrigation application losses	25
2.	Reduction in labour requirements	35
3.	Enhancement in irrigated area	2
4.	Increase in crop yield	20

Zero and Reduced Tillage

Intensive soil tillage is the main cause of reduced soil organic matter and hence of soil degradation. Tillage accelerates the mineralization of organic matter and destroys the habitat of the soil life. On the contrary, when soil tillage is reduced or eliminated, soil life returns and the mineralization of soil organic matter slows down, resulting in better soil structure. Under zero tillage the mineralization of soil organic matter can be reduced to levels inferior to the input, converting the soil into a carbon sink (Reicosky, 2001). Zero tillage also results in water saving and improved water-use efficiency: since the soil is not exposed through tillage, the unproductive evaporation of water is reduced while water infiltration is facilitated. The potential water saving through zero tillage varies according to the cropping system and the climatic conditions. On average, water savings of 15 to 20 per cent can be expected (PDCSR, 2005). Used in isolation, however, zero tillage can lead to problems with weed control, compaction or surface crusting, depending on the soil type.

Table 13.3: Wheat and Rice Yields under Laser Land Leveling (LLL) and Traditional Land Leveling (TLL) in Haryana and Punjab Determined 1 Year after Leveling

State	*Crop*	*Average Yield (kg/ha)*		*Average Yield Difference (kg/ha)*
		LLL	*TLL*	
Haryana	Wheat	4576	4291	285
	Rice	5617	5295	322
Punjab	Wheat	4444	4083	361
	Rice	6168	5807	361

Table 13.4: Geographic Sistribution of the Rice-Wheat System and Estimated Zero and Reduced Tillage Areas in the Indo-Gangetic Plains of India

States	*Area under R-W System ('000 ha)*	*Area with Zero/Reduced Tillage ('000 ha)*		
		2001-02	*2002-03*	*2003-04*
Punjab	2,190	20	50	215
Haryana	910	97	275	350
Uttar Pradesh	5,130	12.6	45	235
Bihar	1,830	0.4	1	18
West Bengal	330	0	0	0
Total Area	**10,400**	**130**	**371**	**818**

One of main reasons for low yields of wheat in rice-wheat cropping system is delayed planting of wheat due to late maturing of preceding rice crop sown in region besides high cost of land preparations and other inputs. After rice harvest, sufficient residual moisture is generally available to establish new crop. Conventional tillage accelerates soil moisture evaporation and requires extra irrigation water to bring field back to semblance of a seedbed. This causes major delays in wheat sowing, which ultimately affects final crop yields. Decrease in wheat yield @ one per cent per day after mid November is well documented due to delay in sowing. Zero tillage is an innovation that not only offers conservation of water and energy resources but also results in better crop yields. This technology had been in use since long in many parts of world and then it was introduced in India. Zero tillage technology has been rapidly accepted by farmers due to its contribution in reducing cost of production, conservation of resources, and improving yields. Malik *et al.* (2005a) was in view that a profit-driven advantage of zero tillage technology has allowed small and medium farmers to gain confidence in this technology.

Bed and Furrow Planting

It refers to a cropping system where the crops grown on beds and the irrigation water are applied in furrows between the beds. This is common practice for the row crops, but not for small grain crops such as wheat and rice. The technique offers a number of advantages, such as improved fertilizer efficiency, better weed control and reduced seed rate. It also saves irrigation water compared to a flat inundated field: the evaporation surfaces reduced and water application and distribution efficiency are increased. In addition, the rooting environments changed and the aeration of the bed zone is better than in flat planting. Reported water savings (compared to flat surfaces) reach 26 per cent for wheat and 42 per cent for transplanted rice, and yield increases 6.4 per cent for wheat and 6.2 per cent for rice (RWC-CIMMYT, 2003).

Bed and furrow planting technologies permit growing of crops on beds with less water. This technique has been tested for various crops and proved quite successful for wheat, maize, rice, *etc.* Advantages associated with bed and furrow technology of crop production are given below:

- ✰ Saving of about 30 per cent irrigation water.
- ✰ Less reduced chances of plant submergence due to excessive rain or over-irrigation.
- ✰ Lesser crusting of soil around plants and, therefore, more suitable for saline and sodic soils.
- ✰ Adaptable for various crops without changing basic design/layout of farm.
- ✰ Enhanced fertilizer use efficiency due to local application.
- ✰ Minimum chances of crop lodging

Direct Seeding of Rice

International Rice Research Institute (IRRI) developed the "aerobic rice technology" to address the water crisis problem in tropical agriculture. Anaerobic rice systems, rice is grown like an upland crop with adequate inputs and supplementary irrigation when rainfall is insufficient (Bouman and Tong 2001). This concept of aerobic rice may be an alternate strategy, which combines the characteristics of both upland varieties with less water requirement and irrigated varieties with high response to inputs. The water use for aerobic rice production was 55-56 per cent lower than the flooded rice, with 16-19 times higher water productivity and net returns to water use was two times higher. The water productivity in aerobic rice is ranged from 0.45 - 0.55 g grain/liter of applied water as compared to 0.25-0.30 g grain/liter of applied water in conventional system. The results of aerobic rice indicated that it may be a viable option where shortage of water does not allow growing lowland rice. Several technologies have been developed to reduce water loss and increase the water productivity of the rice crop; these include the following practices such as saturated soil culture (Borell *et al.*, 1997), alternate wetting and drying (Tabbal *et al.*, 2002), ground cover systems (Lin *et al.*, 2002), and system of rice intensification (Stoop *et al.*, 2002). However, fields are still kept flooded for some periods in most of these systems, so water losses remain high. Aerobic rice is high yielding rice grown under non-flooded conditions in non-puddled and unsaturated (aerobic) soil. It is responsive to high inputs, can be rainfed or irrigated, and tolerates (occasional) flooding (Bouman and Tuong, 2001).

Direct seeding of rice – compared with transplanting – may be considered an RCT:

- ✰ It saves labour and fuel.
- ✰ Seeding into dry soil saves water as there is no puddling.
- ✰ The total growing period from seed to seed is reduced by about 10 days.
- ✰ Yields and water efficiency of the subsequent rotation crops are increased (PDCSR, 2005).

On the other hand, weed management is more difficult in dry direct-seeded rice than in puddled and transplanted rice (RWC-CIMMYT, 2003).

System of Rice Intensification

The system of rice intensification was developed in Madagascar by Fr Henri de Lau Lanie in association with NGO- Association Tefy Saina (ATS) and many small farmers in the 1980s is becoming popular in many countries including India. SRI is a system methodology (Uphoff, 1999; Ghosh *et al.*, 2014) rather than a technology. It is based on the insight that rice has the potential to produce more tillers and early transplanting along with optimal growth condition like wide spacing, optimum humidity, a vibrant healthy soil and aerobic soil conditions during vegetative growth can fulfill this potential. Water saving in SRI maybe as high as 40 per cent as compared to conventional practice. In a field trial at Directorate of Rice Research (Hyderabad, India), SRI gave 166 per cent higher grain yield than normal transplanting method. The varietal response to SRI and normal cultivation was wide SRI method gave nearly 46 to 48 per cent higher yield in hybrids, 52 to 17 per cent in HYVs while negative results were also observed in case of Pusa Basmati due to its shy tillering habit under wider spacing. All the varieties are not promising for SRI cultivation method and response of cultivars to SRI varies as per their ability to exploit the natural resources. Hence, there is a need to develop varieties that can give better response to SRI cultivation and must have compact plant type, profuse tillering, better root system, bolder grains, low water requirement, responsive to organic inputs (inorganic inputs constitute 25 to 50 per cent only), and resistance to pest and diseases. Besides, rice matures 10 – 15 days earlier as compared to conventional practice and thereby vacates the land for timely sowing of succeeding crop. Therefore, genotypes used for SRI should be able to produce more with less duration. High yield varieties and hybrids are the most suitable cultivar for system of rice intensification. In addition, high tillering rice cultivars are also recommended for SRI. Research work conducted at BCKV (Ghosh *et al.*, 2014) during 2012-14 revealed that in Weed management in comparison to weedy check the hand weeding, chemical herbicides and botanical aqueous extracts from

Parthenium hysterophorus, *Tectona grandis* and *Calotropis procera* followed by two mechanical weeding (paddy weeder) at 20 and 40 DAT yielded 27.97, 29.39 and 23.67 per cent more, respectively. Average yield increase in 2014 than 2012 was 13.5 per cent. The major outcome is farmers are accepting gradually the low cost weed management and self-dependent – not depending fully on labours for weed control. In Nutrient management the treatment INM - 50 per cent ORM and 50 per cent INORF (replacing of 50 per cent inorganic fertilizers through organic manures like compost *etc.*) resulted more yield in 2013, but in the final year 2014 the treatment INM 75 per cent ORM and 25 per cent INORF (replacing of 75 per cent inorganic fertilizer by organic manures) resulted better yield, lesser pest attack and better soil health status. Average yield increase in 2014 than 2012 was 11.9 per cent. In Water management average yield increase in 2014 than 2012 was 11.49 per cent. Irrigation at Soil hair crake stage during vegetative phase + 2- 3 cm of standing water at PI and Flowering stage only (SRI – 2 to 3 cm WS at AT, PI and F +HC) resulted 32.78 per cent water savings in comparison to farmers practice – the traditional 3-5 cm water submergence. The other treatments irrigating at hair crack stage (HC) and continuous 2-3 cm water submergence showed 53.9 and 21.23 per cent water saving over farmers practice, respectively. The grain yield data revealed that SRI methodology recorded 15.66 per cent more yield in the final year 2014 in comparison to initial year 2012 and the corresponding figures for the other two treatments [irrigating at hair crack stage (HC) and continuous 2-3 cm water submergence] were 4.10 and 13.08 per cent indicating SRI methodology is better. For producing one kg of paddy the water required 1184, 1075 and 1550 liters of water in SRI, HC and 2-3 cm WS, respectively which are lower than 1891.11 liters of water that required in FP (3-5 cm WS). The major two important greenhouse gases (GHGs) – Nitrous oxide (N_2O) emission ranged from 16-25 kg ha^{-1} and Methane (CH_4) ranged from 25-32 kg ha^{-1} in SRI methodology whereas in FP (3-5 cm WS) the Nitrous oxide (N_2O) emission was 17-21 kg ha^{-1} and Methane (CH_4) ranged from 40-46 kg ha^{-1}. This showed that Methane emission was reduced by irrigating in SRI methodology.

Mulching and Green Manure

The supply of organic matter to the soil through mulching and green manure is important for maintaining and enhancing soil fertility. Mulching material can come from crop residues or green manure crops; it provides feed for the soil life and mineral nutrients for the plants. If legume crops are used as green manure they can supply up to 200 kg/ha of nitrogen to the soil; in the case of rice, this can result in mineral fertilizer savings of 50 to 75 per cent (RWC-CIMMYT, 2003). The spreading of mulch on the soil surface reduces evaporation, saves water, protects from wind and water erosion, and suppresses weed growth.

Controlled Traffic Farming

Controlled traffic farming restricts any traffic in the field to the same tracks. While these tracks become heavily compacted, the rooting zone does not at all, resulting in better soil structure and higher yields. The area lost in the traffic zones is easily compensated for by better growth of plants adjacent to the tracks so that overall yields are usually higher than in conventional systems with random traffic

(Kerr, 2001). Controlled traffic farming is the ideal complement to zero tillage or to bed planting systems, but it also provides advantages in conventional agriculture through time and fuel savings, since resistance to soil tillage in the compaction-free rooting zones is significantly lower and traction is more efficient when tyres work on compacted tracks (RWC-CIMMYT, 2003). In the latter case, either GPS (global positioning system) guidance or visible bed and furrow systems must be used to limit the tillage operation to the rooting zones and to not disturb the tracks.

Synergies between Resource Conserving Technologies

Resource-conserving technologies provide scope for synergy, for example, between bed planting and controlled traffic or mulching and zero tillage. Used in isolation, any of these technologies may face specific problems (*e.g.* surface crusting or weeds in direct seeding rice) or have limitations (*e.g.* zero tillage under irrigated conditions). The combination of resource-conserving technologies working in synergy is commonly referred to as "conservation agriculture".

Conclusion

Resource-conserving technologies applied in isolation have advantages and disadvantages; they are not universally applicable as the problems can sometimes outweigh the benefits. However, by combining different resource-conserving technologies, synergies can be created to eliminate the disadvantages of single technologies and accumulate the benefits. Different RCTs are successfully applied under the concept of conservation agriculture in different cropping systems around the world, allowing stable agricultural production without the known negative environmental impact. The Rice-Wheat Consortium of the Indo-Gangetic plains has been instrumental in adapting the concept of conservation agriculture to rice-based cropping systems, resulting in higher yields, greater profitability, enhanced soil fertility and better water-use efficiency – it represents a possible route towards sustainable agricultural production in rice-based systems. High water consumption is a particular concern. In regions where cropping mainly depends on groundwater for irrigation purposes and where the groundwater tables are falling dramatically, such as in the Punjab of India, water-saving technologies might not be sufficient to guarantee sustainability of the cropping systems. The combination of different RCTs –such as mulching, direct seeding and double zero tillage – results not only in water saving but also in increased infiltration rates (and hence the recharge of the aquifer during the monsoon season). Combined resource-conserving technologies applied in conservation agriculture produce benefits for the farming sector, the environment and the general public, and it is therefore important to promote and adopt them. FAO and regional partners, such as the Rice-Wheat Consortium, can play an important role in this process.

REFERENCES

Anonymous (2014). Agricultural Statistics at a Glance, 2014 – Directorate of Economics and Statistics, Ministry of Agriculture Department of Agriculture and Cooperation Government of India.

Borrell, A.K., Garside, A.L. and Fukai, S. (1997). Improving efficiency of water use for irrigated rice in a semi-arid tropical environment. *Field Crops Res.* 52: 231-248.

Bouman, B.A.M. and Tuong, T.P. (2001). Field water management to save water and increase its productivity in irrigated rice. *Agric. Water Manage.* 49: 11-30.

Ekboir, J. (2002). CIMMYT 2000-2001 World Wheat Overview and Outlook: Developing No-Till Packages for Small-Scale Farmers. Mexico, DF: CIMMYT.

FAO, WFP and IFAD. (2012). The State of Food Insecurity in the World 2012. Economic growth is necessary but not sufficient to accelerate reduction of hunger and malnutrition. Rome, FAO.

Ghosh, P. K., Das, A., Saha, R., Kharkrang, E., Tripathy, A. K., Munda, G. C., and Ngachan, S.V. (2010). Conservation agriculture towards achieving food security in north east India. *Current Science.* 99(7): 915-921.

Ghosh, R. K., Sentharagai, S. and Shamurailatpam, D. (2014). SRI – A methodology for substantially raising rice productivity by using farmers' improve thinking and practice with farmers' available resources. *Journal of Crop and Weed.* 10(2): 4-9.

Jat, M.L., Chandna, P., Gupta, R., Sharma, S.K. and Gill, M.A. (2006). *Laser land levelling: a precursor technology for resource conservation.* Rice-Wheat Consortium Technical Bulletin Series No. 7. New Delhi, India, Rice- Wheat Consortium for the Indo-Gangetic Plains. 48 pp.

Kassam, A. H. and Friedrich, T. (2009). Perspectives on Nutrient Management in Conservation Agriculture. Invited paper, IV World Congress on Conservation Agriculture, 4-7 February 2009, New Delhi, India.

Kerr, P. (2001). Controlled traffic farming at the farm level. GRDC Research Update, Finley NSW, Australia.

Lal, R. (2013). Climate-Resilient Agriculture and Soil Organic Carbon. *Indian Journal of Agronomy.* 58 (4): 440-450.

Lin, S., Dittert, K., Tao, H., Kreye, C., Xu, Y., Shen, Q., Fan, X. and Sattelmacher, B. (2003). The ground-cover rice production system (GCRPS): A successful new approach to save water and increase nitrogen fertilizer efficiency. *In*: Bouman, B.A.M., Hengsdijk, H., Hardy, B., Bindraban, P.S., Tuong, T.P. and Ladha, J.K. (*eds.*) *Water-Wise Rice Production.* Proceedings of a Thematic Workshop on Water-Wise Rice Production, 8–11 April 2002, International Rice Research Institute, Los Baños, Philippines.

Malik, R. K., Gupta, R. K., Yadav, A., Sardana, P. K., Punia, S. S, Malik, R. S., Singh, S. and Singh, S. (2005a). The socio-economic impact of zero-tillage in rice-wheat cropping system of Indo-Gangetic Plains, Zero Tillage – The voice of Framers, Technical Bulletin No. 9. Directorate of Extension Education, C.C.S. Haryana Agricultural University, Hisar, Haryana, India.

Malik, R.K., Yadav, A. and Singh, S. (2005b). Resource Conservation Technologies in Rice-wheat cropping system of Indo-Gangetic Plains. *In*: *Conservation Agriculture – Status and Prospects. Abrol, I.P., Gupta, R.K. and Malik, R.K.* (2005).

Centre for Advancement of Sustainable Agriculture, NASC Complex, New Delhi, pp. 13-22.

PDCSR. (2005). *Project directorate for cropping systems research.* Annual Report 2004–05. India, Modipuram- Meerut. 143 pp.

Reicosky, D.C. (2001). Conservation agriculture: global environmental benefits of soil carbon management. *In* L. Garcia Torres, J. Benites and A. Martinez-Vilela, eds. *Conservation agriculture, a worldwide challenge,* Vol. I, p. 3–12. Rome, FAO and Brussles, ECAF.

RWC-CIMMYT. (2003). *Addressing resource conservation issues in rice-wheat systems of South Asia: a resource book.* New Delhi, India, Rice-Wheat Consortium for the Indo-Gangetic Plains – International Maize and Wheat Improvement Centre, 305 pp.

Sayre, K. (2000). Effects of tillage, crop residue retention and nitrogen management on the performance of bed-planted and furrow irrigated spring wheat in northwest Mexico. Paper presented at the Conference of the International Soil Tillage Research Organization, 15; Fort Worth, Texas, USA; 2-7 Jul, 2000.

Stoop, W.A., Uphoff, N. and Kassam, A. (2002). A review of agricultural research issues raised by the system of rice intensification (SRI) from Madagascar: Opportunities for improving farming systems for resource-poor farmers. *Agric. Syst.* 71: 249-274.

Tabbal, D.F., Bouman, B.A.M., Bhuiyan, S.I., Sibayan, E.B. and Sattar, M.A. (2002). On-farm strategies for reducing water input in irrigated rice: A case studies in the Philippines. *Agric. Water Manage.* 56: 93-112.

Uphoff, N. (1999).Agroecological implications of the System of Rice Intensification. *Environment, Development and Sustainability,* 1: 297-313.

Yadav, J.P.S. and Singh, G.B. (2000). Natural Resource Management for Agriculture production. International Conference on Managing Natural Resources for Sustainable Agriculture Production in the 21st Century: Extended Summaries: New Delhi, India, February 14-18, 2000, Indian Association of Soil and Water Conservationists.

Conservation Agriculture Technologies *Pages* **161–167**
Editor: **Rahul Kumar, Abhinandan Singh and Pankaj Kumar Ojha**
Published by: **BIOTECH BOOKS, NEW DELHI**

14 Conservation and Production of Fodder Crops for Higher and Sustainable Milk Production

A.V. Dahiphale, Sandeep Kumar, Hari Singh and S.K. Kashyap

Department of Agronomy, Institute of Agricultural Sciences, Banaras Hindu University, Varanasi – 221 005, Uttar Pradesh

Adequate attention to forages was given on fodder acceleration and is being more vigorously pursued through feed and fodder related research. The gap in feed and forage supply can effectively be reduced through appropriate crop production management options. Efficient forage production strategy need to follow to ensure sustainability, high level of productivity and economic viability. Quality milk production only can achieved by site specific fodder crop production system. Fodder productivity can be enhanced through focusing on new options viz. forages for alternate land use, problem soils, farming system and new niches of forage production in different agro-climatic regions. Intensive forage production systems under rainfed situation for round the year forage production system is a key for minimizing demand and supply of the fodder production system in country.

Introduction

In India out of total geographical area of 329 m ha an area of 12.01 m ha has been classified as grasslands (pastures and other grazing lands). These are cultivable lands, being used as permanent pastures and meadows in hill slopes, and village common lands in plains. High altitude pastures are relatively less affected by erosion; the plane village grazing lands are severely eroded. About 40 per cent *i.e.*, 4.2 m ha of total area has estimated as affected by erosion (Das, 1985). The common practice of feeding milch cows and buffaloes is to feed them on crop residues with little supplementation of green forages (if available) and concentrates. But concentrates feeding results in increased cost of feeding. Feeding green forages, however, reduces the cost and has the potential of higher level of milk production. Nearly 65 per cent of the total expenditure of milk production in crossbred cattle is attributed to the feeding of animals when both concentrates and green fodders were fed as mixed ration. But when the milk production depend upon concentrate based feeding, the cost of feeding towards milk production was 83 per cent while on forage based feeding, it could be reduced to only 40 per cent of the total expenditure. Although, milch animals specially require good quality forages (which have many additional benefits) for expression of full genetic potential of milk production. Intensive forage production systems aim at efficient utilization of land and other inputs for higher fodder production per unit area per unit time with equal quantum of increase in animal productivity. These systems include multiple cropping, intercropping, overlapping cropping and relay cropping. The system of multiple cropping ensures regular supply of forage by manipulating the sowing and harvesting schedules. These crop sequences are tai-lored with an objective of achieving high yield of green nutritious forage and maintaining the fertility of the soil (Hazra, 1988; Sunil Kumar, 2007). In humid and sub-humid climate, setaria and guinea grass perform better. Berseem can be replaced by lucerne in western part of the country for utilization of interspaces. Except for high water table areas, rotations based on perennial crops are most suitable, whereas in other cases annual crops in rotation are suitable. Overlapping system is suitable for large dairy farmers and food and forage based system is better for small farmers having limited land holdings.

Economical Food-Fodder Based Cropping Systems

Efficient forage based cropping systems with respect to biological potential along with increased efficiencies of land and water may provide a viable option for specific regions. At farmer's level, productivity potential and mon-etary benefits act as guiding principles, while opting for a particular crop/cropping system. Introduc-ing the leguminous fodder crops in wider spacing of grain sorghum is an excellent example of food-fodder cropping systems. Some of the examples are sorghum/ groundnut – berseem – maize + cowpea (Bundelkhand region), rice-berseem – forage maize + cowpea (Tarai), maize + cow-pea- lucerne (central Maharashtra).

Forage Based Cropping in Rainfed Areas

Since last 4-5 decades, rainfed forage agronomy has been oriented towards identification of suitable crops and their varieties, possibility of integrating forages in the existing cropping system, resource management and contin-gent crop planning. In rainfed areas, integrated system of food - fodder production such as intercropping, ratooning and over seeding of grain crops have demonstrated to mitigate shortage of feed and fodder. The productivity of predominantly mono cropped areas could be raised through intercropping of short duration forage crops in long duration rainy season food crops like grain sorghum and pigeonpea. For mitigat-ing occasional long droughts and supply of regular fodder, in western and southern regions of the country, where sorghum is grown both during *kharif* and *rabi,* ratooning of *kharif* sown crops produces good yield. The combination of graminaceous and leguminous fodder crops improves the herbage quality substantially in terms of protein and mineral balances as the legume component contains higher amount of protein, calcium and phosphorus. The inclusion of bushes brings in the climate resilience in ad-dition to saving in energy, otherwise used in repeated field preparation and sowing (Agarwal *et al.*, 2007). In risk prone drought affected areas, grass and perennial compo-nents based approaches, *viz.* ley farming and alley crop-ping systems hold promise to increase the overall produc-tivity of marginal and sub-marginal lands and strengthen the livestock enterprise and conserve the land-water re-sources. perennial grasses like hybrid napier, Trispecific hybrids (TSH) and guinea grass with bushes like sesbania, subabool has been tested in different planting arrangement alongwith food crops like sorghum, pigeonpea for produc-tion of grain and fodder both. TSH based system has been found more drought tolerant.

The nutrient requirement of for-ages under multi-cut system is much higher than under single system. The work on nutrition to forages with reference to secondary nutrients like Ca and Mg is meager, however with emerging multi-nutrient deficiencies in specific areas, NPK along with S, Zn and Mo based nutrient management has been attempted to correct the deficiency and balancing the nutrients for quality fodder and live-stock health (Sunil Kumar and Faruqui, 2010).

Feeding of Milch Cows/Buffaloes

Feeding of leguminous fodder like green berseem or lucerne and cowpea during winter and summer/rainy seasons, respectively along with 1-2 kg of dry roughage to lactating cows and buffaloes sustained a milk yield upto 10 kg/day. But animals yielding more than 10 kg/day required supplementation of energy rich concentrates @ 1 kg for every 2.5 - 3.0 kg of extra milk. Although, feeding of leguminous fodder exclusively resulted in wastage of protein, the most deficient nutrient in cereal fodder based feeding regime. But a mixture of leguminous and cereal fodder at the ratio of 2: 1 or 1: 1 were found to be more promising. A balanced fodder mixture can be produced by growing legume fodder *viz.*, cowpea, berseem, lucerne, stylo, dolichos, siratro *etc.* and cereal fodder *viz.*, maize, oat, sorghum, hybrid napier, guinea, paragrass *etc.* Cereals like maize/sorghum can be cultivated along with legumes like cowpea during kharif, and berseem and oat during rabi season to supply a balanced fodder mixture. Cows consuming fodder mixture containing one part of berseem and oat each (fresh weight basis) with 1 kg concentrate (for letting down of milk) yielded 7-8 kg of milk/head/day. Similarly, hybrid napier and lucerne (3: 1) or maize and cowpea (1: 1) mixture sustained upto 7 kg milk/ head/day. Thus, the intake and utilization of nutrients from comparatively low quality fodder were improved by judicious inclusion of legumes.

Cereal fodders including grasses are poor in quality, having low CP and high fibre contents coupled with low voluntary intake. Therefore, performance of lactating animals fed on such fodders like maize, sorghum, guinea and hybrid napier grass *etc.* alone, were adversely affected which could meet only the maintenance with little production needs of animals. But supplementation of concentrates improved the level of milk production. However, oat at early stage of growth (upto 10 per cent flowering) sustained 6-8 kg milk yield daily.

Again the surplus green forages available during the flush period of growth can be conserved as hay or silage to provide some quality forage to milch animals during the scarcity period of October-November or May-June. Berseem/Lucerne hay with I kg of concentrates sustained upto 10 kg daily milk yield during summer months of May-June,when there is complete scarcity of cultivated green forage. Although, milk yield was low (5 kg/head/day) in cows fed silages of jowar and cowpea (2: 1) or oat and berseem (1: 1) without any concentrate supplements. Hence, there is a need of supplementing quality concentrate also, in silage based diets of milch animals to sustain their optimum milk yield under hot dry condition of summer months.

Table 14.1: Performance of Milch Cows/Buffaloes Fed all Forage/Forage Based Rations

Sl.No.	*Kind of Forage**	*Level of Supplementation (Head/d)*	*Milk Yield Sustained (kg/d)*
1	Berseem	1-2 kg dry roughage	8-10
2	Berseem	Concentrate @ 1 kg/2.5-3.0 kg extra milk	> 10
3	Cowpea	1-2 kg dry roughage	6-8
4	Cowpea	Concentrate	> 8
5	Hybrid napier + Lucerne (3:1)	1-2 kg dry roughage	6-7
6	Maize + Cowpea (1:1)	1-2 kg dry roughage	6-7
7	Berseem + oat (1:1)	1-2 kg dry roughage	7-8
8	Oat (10 per cent flowering)	1-2 kg dry roughage	7-8
9	Guinea grass	1-2 kg dry roughage	> 5
10	Maize/sorghum/bajra/hybrid napier	1 kg concentrate	4-6
11	Sorghum + Cowpea (2:1) silage	Nil	> 4
12	Oat + berseem (1:1) silage	Nil	5-6
13	Berseem/Lucerne hay	1 kg concentrate (barley)	> 10
14	Cowpea hay (at 50 per cent flowering)	1 kg concentrate (barley)	8-10

* Offered *ad libitum.*

They may be given a ration (Table 14.2) Comprising of little quantity of straws/ stovers, mixture of legume and non-legume (cereal) green forage and good quality concentrate mixture (20 per cent CP and 72 per cent TDN). While cows yielding >21kg milk per day, needs special attention to meet their nutritional requirement. Such high yielding cows will always be deficit in energy in first two trimesters of lactation on conventional diets

Concentrate Mixture

Anybody desirous of producing concentrate mixture of his own should know the various types of feed ingredients required for making concentrate mixture ideal for animal feeding. If he is not well conversant with the quality of raw feed ingredients and unconventional feedstuffs available in the region, he may not be able to formulate an ideal concentrate mixture for his animals economically.

Table 14.2: Rationing (as such basis) of High Yielding Milch Cows

	Feedstuffs/Items	*Quantity (kg/head/day)*
A. Category-I (350-400 kg b.wt, 15 kg daily milk yield and 4-5 per cent milk fat)		
1.	Maize/Sorghum/Oat	20
2.	Berseem/Lucerne/Cowpea	25
3	Wheat/Rice straw	1
4.	Concentrate mixture	5
B. Category-II (350-400 kg b.wt, 25 kg daily milk yield and 4-5 per cent milk fat)		
1.	Maize/Sorghum/Oat	20-25
2.	Berseem/Lucerne/Cowpea	20-25
3.	Concentrate mixture	8-9

Conclusion

In climate changing agricultural scenario, need based site specific options should be identified for making forage production to meet demand of green and dry fodder. Resource optimization for increased input use efficiency in different farming situations, forage production from problem soil, Forage production for lean and scarcity period. Intensive forage production systems under rainfed situation for round the year forage production with different land holdings having emphasis on pro-longed availability of leguminous fodder feeding system of fodder crop is directly responsible for the quality milk production. It can only be enhanced by improving the proper planning of fodder components in the existing crop production system.

REFERENCES

Agrawal, R.K., Niranjan, K.P. and Rai, S.K. 2007. Evaluation of perennial grasses based cropping system for prolonging for-age productivity under semi-arid rainfed situation. *Range Management* and *Agroforestry* **28** (2B): 386-387.

Arora, S.P. 1978. Feeding of Dairy Cattle and Buffaloes. ICAR, New Delhi.

Banerjee, G.C. 1988. Feeds and Principles of Animal Nutrition. Oxford and IBH Publishing Co. Pvt. Ltd., New Delhi. *Dairy Sci.* 64: 1-13.*Dairy Technol.* 14: 207-215.

Das, D.C. (1985) Problems of Soil Erosion and Land Degradation in India. National Seminar on Soil Conservation and Watershed Management, September 17-18, New Delhi.

Hazra, C.R. 1988. Forage production on problem soils. In pasture and forage crops research. A state of knowledge report (Punjab Singh ed), RMSI, IGFRI, Jhansi, pp. 230-238.

Jain, M.K., Sunil Kumar and Yadav, S.S. 2007. Influence of preced-ing crops on performance of *rabi* food and forage crops un-der irrigated condition of Bundelkhand. *Forage Research* **33** (1): 70-72.Ltd., New Delhi.

Mahanta, S.K. and Pachauri, V.C. 1999. Importance of quality forage to milch animals for economical milk production. Milcow 21: 26-28.

Mahanta, S.K., Singh, K.K, Das, M.M and Das, N. 2009. Forage based feeding of livestock. In: Forage for Sustainable Livestock Production (eds. N. Das, A.K. Misra, S.B. Maity, K.K. Singh and M.M. Das). Satish Serial Publishing House, Delhi, India pp. 407-426.

Mertens, D.R. 2009. Maximizing forage use by dairy cows. *Adv. Dairy Technol.* 21: 303-319.

Miller, W.J. 1979. Dairy cattle Feeding and Nutrition. Academic Press, London.

Mott, G.O. and Moore, J.E. 1969. Forage evaluation techniques in perspectives. In: R.F.

Barness, D.C. Clauton, C.H. Gordon, T.J. Kloptenstein and D.R. Waldo (eds.), Proceed. Natl. Conf. On Forage Quality Evaluation and Utilization. Nebraska Cent. Educ., Lincoln, NE.

Mudgal, V.D. 1992. Formulation of economic rations for dairy cattle under field conditions. Proceed. All India dairy husbandry officers' workshop, Oct. 22-23, 1992 at NDRI, Karnal.

Pathak, N.N. and Jakhmola, R.C. 1983. Forages and Livestock Production. Vikas Publishing House Pvt. Ltd., New Delhi

Pathak, P. S. and Roy, M.M. 1995. Agri-silvipastoral farming sys-tems for optimizing forage and energy resources in rainfed areas *In.* Forage Production and Utilization (Eds. R. P. Singh), Indian Grassland and Fodder Research Institute, Jhansi pp.154-178.

Pathak, P. S. and Sunil Kumar. 2004. Forage and grazing resources in different agro-climatic regions in India (in): *Buffalo pro-duction under different climatic regions,* Kundu S. S., Mishra A K and Pathak P S (Eds.). International Book Dis-tribution Co. Lucknow. pp. 1-41.

Pathak, P.S. and Ram Newaj, 2003. Agroforestry: Potentials and Opportunities, Agrobios (India), Jodhpur.pp-1-337.

Pathak, P.S. and Solanki, K.R. 2002. Agroforestry technologies for different agro-climatic regions of India, Natural Resource Management Division, ICAR, New Delhi p. 42.

Prasad, L.K. 1995. Sustainable forage production systems in East-ern plateaus of India. In. New vistas in forage production (Eds) Hazra, C.R. and Misri, Bimal, IGFRI, Jhansi: 113-119.

Prasad, N.K. 2002. Fifty years of forage and agroforestry research, Directorate of Research, Birsa Agricultural University, Ranchi, Jharkhand, p. 53.

Ranjhan, S.K. 1993. Animal Nutrition in the Tropics. 3rd rev. edn. Vikas Publishing Pvt., New Delhi.

Stokes, S. 2002. The importance of forage quality for milk production and health. *Adv.*

Stone, B. 2011. High quality forages: more milk, less grain. http: / /www.diamondv. com/newscenter_documents/NutritionLineFeatureArticles/Dairy20 10/ May2010.

Sunil Kumar and Faruqui, S. A. 2010. Forage production technolo-gies for different agro-ecological regions, *Tech. Pub. No. 01/2010, AICRP on Forage Crops, IGFRI, Jhansi*, p. 64.

Sunil Kumar. 2007. Advances in forage production technology: An Indian perspective In: *Special Summer School Training on Recent Advances in Agricultural Sciences* on 23 May, 2007 CSA UAT, Kanpur. pp: 125-140.

Thakur, S.S. 2001. Scientific feeding for higher milk production. In Short Course on 'Biotechniques to augment lactation in bovines, May 22-31, 2001 at NDRI, Karnal.

Wangsness, P.J. and Muller, L.D. 1981. Maximum forage for dairy cows: review. *J.*

Conservation Agriculture Technologies *Pages* **169–182**
Editor: **Rahul Kumar, Abhinandan Singh and Pankaj Kumar Ojha**
Published by: **BIOTECH BOOKS, NEW DELHI**

15 Nematode Responsive Promoters for Protecting Crop from Nematode Attack

Ashish Kumar Singh[1], Ila Joshi[2], Neeraj[1], Rahul Kumar Tiwari[3] and Animesh Kumar

[1]Division of Nematology, [2]NRCPB, [3]Division of Plant Pathology,
ICAR-Indian Agricultural Research Institute, New Delhi – 110 012
Central for Agriculture Bioinformatics, New Delhi

Plant parasitic nematodes (PPNs) are recognized as a severe pest of agricultural crops. Loss to world agriculture because of plant parasitic nematodes is estimated at about 173 billion US$. Most of the damage to agricultural crops is by sedentary plant parasitic nematodes namely, root-knot nematodes (Meloidogyne spp.), cyst nematodes (Heterodera and Globodera spp.)and reniform nematode (Rotylenchulus reniformis).As per world bank report up to 35 per cent increase in population is estimated by the year 2050 and it will be accompanied by an increase in food demand by 75 per cent. Management of these ubiquitous pests is a major concern as the existing management options are not suffice enough to prevent the economic losses rendered by nematodes. Genetic engineering provides several novel strategies for nematode management which are safe and environment friendly. But for developing these transgenic strategies, we need a promoter to help express the genes. A nematode responsive and root expressive promoter is an ideal choice as compared to any systemic expressing promoter.

Keywords: *PPNS, Root knot nematodes, Cyst nematodes and Reniform nematodes and Promoters.*

1. Introduction

Nematodes are microscopic, cylindrical shape, unsegmented, bilaterally symmetrical and pseudocoelomatic invertebrates and need a thin film of moisture for active life. Nematodes are most copious metazoans on earth (Ferris *et al.*, 2001) as they havesuccessfully colonized a variety of habitats, be it sea, mountains, desertsor plains. The phylum Nematodaincludesmore than 25000 species (Blaxter 2003). Nematode can be grouped into free-living and parasitic nematodes; where in the former group inhabit soil, freshwater, and marine environment and play a crucial role in nutrient recycling while the latter group comprises of parasites of animal and plants. Approximately 4100 species of plant parasitic nematodes (PPNs)have been recognized (Decraemer and Hunt 2006) and they inflict damage of about 173 billion dollars to the world agriculture annually (Elling, 2013). All plant species harbour one or more nematode species which feed either ecto-parasitically or endo-parasitically on root or shoot of plants. However, despite of their detrimental effect on growth of crop there is a lack of manifestation of any specific above ground symptom and as a result it becomes very difficult to convince farmers, growers, advisors *etc.*of the possible damage by plant parasitic nematodes. Mostly thesymptoms are similar to that of nutritional deficiency,encouragingfarmers to apply fertilizers, thus increasing the cost of cultivation and soil pollution. PPNs are a major global challenge to food security of the world and this necessitatesdevelopment of effectivestrategies for their management and control.

A wide range of strategies to manage nematodes are being utilized but provide limited success. Conventional methods largely rely on use of nematicides which impose high financial burden on growers and drastically affect environmentand human health. As a result, most of the chemicalnematicides have been withdrawn by many countries or are undergoing a phased withdrawal under the UN Montreal protocols. Use of bioagents for nematode management is still in development process and not viable for large acreage. Another major approach which is farmer and environment friendly is the use of natural resistance (Cook and Evans, 1987; Atkinson, 1995). Robert (1922) have reviewed both success and limitation of this approach, mostly problems are centred on availability of vertical resistance, long breeding period, durability under different genetic variability of pathogen.

A major hope for future is engineering plant resistance which is a novel, time efficient, cost effective, ecofriendly, durable, and sustainable strategy for limiting crop damage due to PPNs. RecombinantDNA technology offers us avenues to transfer putative genes across species barriers for engineering resistant crop varieties. *In planta*RNAi strategies targeted at crucial genes of nematodes, offer good hope for engineering resistance against the nematodes. But in all these cases the gene expression should be limited to the root system (zone of nematode feeding) and would be further better if the gene expression is specific to nematode penetration. The ability to drive gene expression in precise spatio-temporal manner is an important factor and can be achieved by carefully selecting the promoterregions used in gene constructs. Promoter is an upstream region (towards5′ end) of DNA molecule where RNA polymerase binds and initiates transcription of a particular gene.In other words promoter is a regulator of gene expression. Till date most of thedsRNA constructs, eliciting RNAi,have been driven through constitutive promoter *CaMV35S*, which keep expressing all the time all across the plant and hence consume huge amount of energy resources. Also in some cases this promoter was observed to be down-regulated in nematode feeding sites (Goddjin *et al.*, 2003).

Hence, a promoter that confines gene expression to roots will be an ideal requirement for developing transgenic resistance against PPNs.

2. Promoter

Nematodes are very difficult to manage. They are at their most vulnerable during their active phase in soil when searching for host. Most of the management strategies, especially physical, cultural, biological are aimed to disrupt nematode development at this stage. Deep summer ploughing exposes the nematode larvae to heat and moisture stress, thus killing the infective larvae. Soil solarisation involves covering of the soil with clear plastic, entrapping the radiation and generation of heat, which kills the nematode larvae in soil. The biological control agents like, the bacterial parasite, *Pasteuria* spp., or nematophagus fungi like *Arthrobotrys* spp., *Monacrosporium* spp. attack or predate on the infective nematode larvae in soil. The chemical fumigants or nematicides applied in soil also target the nematode life cycle stages in soil pre-penetration. Once the endoparasitic nematode species have entered a root, control by even chemicals becomes challenging as nematicidal compounds have to be non-phytotoxic and preferably systemic. A nematicide that can be safely applied to growing plants and is translocated to the roots in sufficiently large amounts to kill endoparasitic or ectoparasitic nematodes has not been discovered. Oxamyl, a systemic compound that is translocated basipetally, is the only commercial product that is used as a foliar treatment, but its use as a liquid formulation is restricted in many countries for toxicological reasons. Most of the fumigants and insecticides being used as nematicides have been banned because of environmental contamination and human toxicity issues.

Natural plant resistance against a particular PPN is the most cost effective, environment as well as farmer friendly management option, but very few crops harbour a resistance gene. With the advent of recombinant technology, genetic engineering allows us to engineer resistance in crop plants against nematode pests by either transferring the resistance gene across crop plants or devise novel approaches which are detrimental to nematode feeding, development and multiplication.

Some of the successful novel approaches found effective are as follows:

2.1 Destruction of Feeding Cells

The endoparasitic PPNs like, the root knot, cyst and reniform and citrus nematodes alter the host physiology and form a specific feeding cell. These altered cells act as "nutrient sink" and provide adequate nourishment for the development of nematodes. Thus an approach which destroys the feeding cells could be a strategy to hamper further development of the nematode. This approach mimics programmed cell death which is central to the elimination of pathogen by natural *R* (Resistance) gene such as *Mi* and *Gro-1*. A plant cell can be damaged expressing a phytotoxic gene product particularly in feeding cells like *barnase* (bacterial RNAse) with the inhibitor protein *barstar* under control of *CaMV35S* promoter. This has been validated against *M. incognita* (Opperman, 1994) and *H. schachtii* (Ohl, 1997) successfully.

2.2 Impairing the Efficiency of Nutrient Supply by Feeding Cells

Modified cells like syncytia, giant cells created by nematodes resemble plant transfer cells. These cells permit them to draw more nutrients from plants. Hence, constraining the nutrient exchange capacity of these modified cells can be considered a potential target. When a water channel protein *TobRb7* was targeted using antisense construct of this gene, it provided partial resistance to plants against *M. incognita* (Opperman, 1996).

2.3 Inhibiting Digestive Proteinases

Serine and cysteine proteinases play a crucial role in assimilation of food material in nematodes. Proteinase inhibitors occur naturally in plants which provide natural defense. When cowpea trypsin inhibitor was expressed in transgenic potato, it influenced the sexual fate of *Globodera pallida*. However, it didn't reduce the fecundity of females (Hepher and Atkinson, 1992; Atkinson, 1993).Conversely; CpTI reduced the fecundity of females of *M. incognita* without affecting their sexual fate (Hepher, 1992; Urwin, 1998).

2.4 Gene Silencing

RNAi has emerged as a potent and successful technology for crop protection in recent years, very specially against nematodes. The *in planta* delivery of dsRNA to feeding nematodes leads to the specific down-regulation of their target genes. The advantage of RNAi based transgenic approach is that there is no synthesis of any gene product. RNAi approaches have been targeted at silencing the nematode parasitism genes, house-keeping genes, developmental genes, *etc.* of critical importance to nematodes. These have been reviewed recently by Dutta *et al.* (2015).

All the above strategies require a robust promoter for the expression of the transgenes. Our work aims at validating a nematode responsive root expressing promoter. A promoter is a DNA sequence flanking the structural gene, and acts as a control sequence where RNA polymerase must bind to initiate transcription of the flanking structural gene into messenger RNA and protein. Plant transformation for development of transgenic plant involves the construction of an expression vector comprising of a heterologous DNA under control of or operatively linked to a regulatory element (for example, a promoter).

Promoters regulate temporal and spatial gene expression through DNA recognition sequences, which interact with basic transcription factors. Such DNA recognition sequences usually include a core promoter with upstream enhancer sequences located close to the structural portion of the gene. Transcription can be activated by these enhancer sequences independent of their location, distance or orientation with respect to the genes promoters (Figure 15.1).

3. Promoters for Nematodes Control

Promoters used in biotechnology applications can be assembled into the following classes: (a) constitutive promoters; (b) spatiotemporal promoters; (c) inducible promoters; and (d) synthetic promoters.

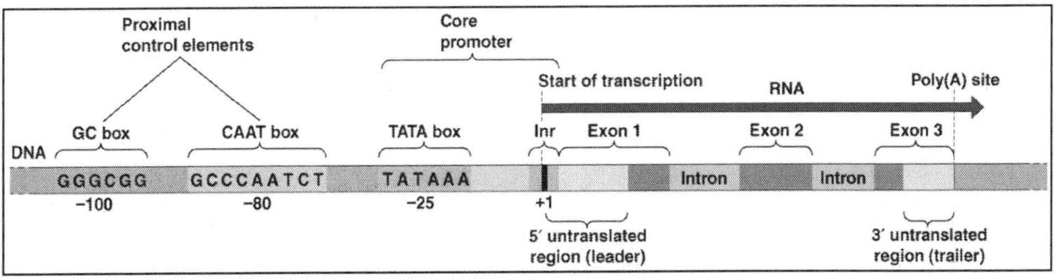

Figure 15.1: Structure of a Promoter (http: //mol-biol4masters.masters.grkraj.org/html/ Gene_Expression_II5B-Mechanism_of_Transcription.htm).

3.1 Constitutive Promoters

Constitutive promoters drive constant levels of gene expression in all tissues, at all times. Among the viral promoters, the *CaMV35S* promoter from the cauliflower mosaic virus is one of the most widely utilized promoters for basic research and the development of transgenic plants.

3.2 Spatiotemporal Promoters

These promoters provide more precise control of native genes and transgenes, and restrict gene expression to certain cells, tissues, organs, or developmental stages. An ideal tissue-specific promoter will only be active in certain tissues leaving all other tissues unaffected.

3.3 Inducible Promoters

These promoters are responsive to environmental stimuli and provide precise regulation of transgene expression through external control. Inducible plant promoters can be usually classified as responsive to endogenous signals (plant hormones), external physical stimuli (biotic and abiotic stresses), and external chemical stimuli. The ideal promoter for designing resistance against nematodes would be the one expressing only in roots and only on nematode infection. NRCPB and our group under a NAIP project have identified a promoter from Arabidopsis whose activity is triggered by *M. incognita* and is restricted to only roots. We intend to see its activity in response to rice root-knot, pigeon pea cyst and reniform nematodes.

4. Role of Nematode Responsive Promoters in Managing Nematode Populations

Nematode responsive promoters are those whose transcription initiation activity is either induced or repressed in response to a nematode stimulus. Thus, a nematode inducible promoter increases expression of respective gene in the presence of a nematode stimulus. In contrast, a nematode repressible promoter decreases the transcription of a gene in the presence of a nematode stimulus. Nematode regulated promoters provide a means for improved regulation of genetically engineered nematode resistance in plants. It is known that expression of a toxin gene product in nematode feeding sites can potentially show its effects on the off target sites. Thus, it can be beneficial to additionally alter the transgenic plant to express a product that counteracts excessive production of the toxin outside of the feeding cell or of the feeding zone.

A nematode repressive promoter can be used in combination with a nematode inducible promoter to effect enhanced regulation of nematode resistance in a plant. In this way, two transgene units in one or two nucleic acid molecules are used in combination to transform plant cells and regenerate transgenic plants having improved nematode resistance. The first transgene unit comprises a nematode inducible promoter operably linked to a nematode resistance sequence. The second nucleic acid molecule comprises a nematode repressible promoter operably linked to a heterologous nucleotide sequence that encodes a gene product that, when expressed in a plant cell inhibits or inactivates a toxic product of a nematode resistance gene that has been engineered within the plant cell.

Promoters can also be applied for RNA interference (RNAi) which is the conserved phenomenon of gene silencing mediated by double-stranded RNA (dsRNA). RNAi is a promising molecular tool with potential application for functional genomics and also control of PPNs. The functional genomics knowledge can be employed to successfully engineer resistance in plants against nematodes. The ability to specifically knock down a selected mRNA transcript allows the investigation of gene function which is crucial for nematode parasitism and interaction with the host. The study of gene function through RNAi is well documented for many PPNs. *In planta* RNAi was successfully employed for two genes – one coding for splicing factor and the other coding for integrase of the root-knot nematode *M. incognita* using *CaMV35S* promoter or a tobacco root-specific promoter (Yadav *et al.*, 2006). *In vitro* and *in vivo* RNA interference approaches have been used to silence the parasitism gene*16D10* in root-knot nematode. Ingestion of *16D10* dsRNA driven by *CaMV35S* promoter *in vitro* has been reported to silence the target parasitism gene in RKN and resulted in reduced nematode infectivity (Huang *et al.*, 2006). Development of resistance to *M. incognita* through host-derived RNAi has been demonstrated for two FMRFamide like peptide genes *flp-14* and *flp-18*. The study demonstrated both *in vitro* and *in planta* validation of RNAi-induced silencing of the aforementioned genes cloned from J2 stage of *M. incognita* driven by *CaMV35S* promoter (Papolu *et al.*, 2013). Similarly, RNA interference (RNAi) construct *pART27* (*16D10i-2*) was introduced into *Arabidopsis thaliana* and potato against *M. chitwoodi* effector gene *Mc16D10L*. This resulted in a significant level of resistance against *M. chitwoodi* in *Arabidopsis* and potato (Dinh *et al.*, 2014).

With increased knowledge of nematode genomics and transcriptomics researchers shall be able to design resistance against nematodes by transgenic approaches, using putative nematode responsive root specific promoter against multiple PPNs.

5 Promoter Analysed so for against Various PPNs

Goddijn *et al.*, 1993 analysed promoters of the *CaMV35S* gene, bacterial nopaline synthase, rooting loci (rol) and T-cyt genes and the plant-derived phenylalanine ammonia-lyase I gene in transgenic tobacco and Arabidopsis by introducing promoter-gusA fusion constructs. Transgenic plants were analysed histochemically for GUS activity in the nematode feeding structures after infection with either *Heterodera schachtii* or *Meloidogyne incognita.* Down regulation of GUS activity was observed in the feeding structures of both the nematodes, thus, concluding that the development of the nematode feeding structures influences expression of a large number of genes.Ehsanpour and Jones (1996) studied the expression pattern of a GUS reporter gene linked to a *Parasponia andersonii* haemoglobin promoter in transgenic tobacco plants after infection with *M. javanica.* GUS activity was analysed at different time points. GUS activity was observed at early stages of infection (0-2 weeks) and became stronger in later stages of infection (3-6 weeks).Hansen *et al.* (1996), developed transgenic potato roots using *Agrobacterium rhizogenes* and a chimeric transgene consisting of the wound-inducible *wun1* gene promoter and the β-glucuronidase (*uidA*) reporter gene. Transgenic hairy roots were inoculated with the white cyst nematode, *Globodera pallida,* and the root-knot nematode *M. javanica.* Strong GUS expression was detected during intracellular migration by *G. pallida* but later on it was limited to the feeding sites. In contrast, in the case of *M. javanica* infection strong GUS intensity was visible in galls.Urwin *et al.* (1997) analyzed expressions of the *CaMV35S* with GFP in feeding sites of beet cyst nematode (*Heterodera schachtii*) and root-knot nematode, *M. incognita.* Initially the GFP levels were constant in cells surrounding the syncytium induced by *H. schachtii,* but were downregulated progressively from 10 to 35 days post-infection within feeding sites. However, in case of *M. incognita the* GFP levels were upregulated.Moller *et al.* (1998) observed the expression of *atao1* gene of *Arabidopsis thaliana* which encodes an extracellular copper amine oxidase. They developed transgenic plants harbouring a GUS reporter gene under the control of the *atao1* promoter and inoculated with the nematodes *Meloidogyne incognita* and *Heterodera schachtii.* No GUS activity was observed during root invasion by either nematode, however, with progressive infection and development of the feeding site GUS activity was first observed in the gall induced by *M. incognita* between 6 and 10 days after inoculation. By contrast, GUS activity was very limited following infection with *H. schachtii,* only being observed between 13 and 20 days after inoculation, in some cells surrounding the syncytium.Goverse *et al.* (1998) investigated the promoter activity of *CaMV35S* and the mannopine synthase *TR22* with the fusion of the green fluorescent protein (GFP) which was expressed in transgenic potato (*Solanum tuberosum*) plants against potato cyst nematode (*Globodera rostochiensis*). Upregulated GFP levels were observed at 4 dpi in case of both the promoters specific to feeding sites. Escobar *et al.* (1999) isolated and characterized a gall specific (giant cell) promoter *LEMMI9* from tomato plants. Transgenic potato plants carrying a *LEMMI9* promoter: GUS fusion were analysed and it was observed to be activated against *M. incognita* induced galls in a heterologous system.Hutangura *et al.* (1999) studied the effect of nematode infection on auxin distribution in developing galls and suggested that nematodes control auxin distribution in the root. Auxin responsive (*GH3*) promoter and three chalcone synthase (*CHS1, CHS2* and *CHS3*) promoters were analysed by fusing with gusA reporter gene in white clover (*Trifolium repens* cv. Haifa). During the initiation of root galls by root-knot nematodes (*M. javanica*) GUS activity of *CHS* was evident 24 h post inoculation around the invading nematode while in *GH3* it was observed in the gall 48–72 h post inoculation.Bertioli *et al.* (1999), studied the 14 subdomain constructs of *CaMV35S* promoter with various 35S-*uidA* incorporated in tobacco plants and assayed for GUS activity after infection by root-knot nematodes (*M. incognita*) and cyst nematodes (*G. tabacum subsp. tabacum*). Nematode-responsive expression for all the constructs was always found stronger and more consistent with the root-knot nematode than the cyst nematode. Puzio *et al.* (2000) identified *pyk20* promoter from *Arabidopsis thaliana*. *A. thaliana* transgenics were generated with thedeletion constructs fused with GUS. The transcription of the gene was shown to be induced upon infection of beet cyst nematode (*H. schachtii*), in roots infected by a fungus-like organism (*Plasmodiophora brassicae*), by plant hormone treatment and by wounding. Bertioli *et al.* (2001) studied two plant resistance genes, *Cf-4* and *Cf-9*, with their corresponding pathogen avirulence genes *Avr4*and *Avr9*. Both the genes were expressed together in transgenic plants using three nematode responsive promoters (4xB5+A, D0.*3TobRB7* and 35S) in different combinations. Positive expression was shown in root-knot nematode infected plants. The *Cf-4* and *Cf-9* genesexpressed in tobacco roots upon nematode infection.Van Pouck*eet al.* (2001), evaluated expression of five nematode-induced promoters (*ARM1, 1164, 728, 25* and *LEMMI9*) and four down regulated promoters (*CaMV35S*, the nopaline synthase promoter (*nos*) and the rooting loci promoters *RolC* and *RolD*) fused with GUS reporter gene in sugar beet hairy roots (transformation done with *Agrobacterium rhizogenes*). The transgenic hairy roots were challenged with *Heterodera schachtii* and subjected to GUS analysis at different time-points (4, 8, 15, 22 dpi). The upregulated promoters (*728, 25, ARM1*) were found to be active at the base of lateral roots and root tips while the *LEMMI9* promoter and the 4 constitutive promoters were active in the entire root. The highest GUS expression for the *ARM1, 1164* and *728* promoter was observed in syncitia at 8 dpi. The *RolC* and *35S* promoter were found to be down regulated at 8 dpi while the *RolD* promoter was down regulated at 15 dpi. However the nos promoter was down regulated at 4 dpi.Karimi *et al.* (2002) analyzed *Atgrp* promoter activity in NFS after fusion with GUS and transformation into Arabidopsis plants. Transgenic Arabidopsis plants were challenged against root-knot nematode (*Meloidogyne incognita*) and the cyst nematode (*Heterodera schachtii*) and GUS assays were performed at different time points after infection. The up-regulation was observed 3 days after inoculation in the feeding sites of both nematodes.A sunflower heat shock protein promoter *Hahsp17.7G4* (encoding for small heat-shock protein having role in embryogenesis and stress responses) was found to be active in giant cells after *M. incognita* infection in transgenic tobacco (Carolina Escobar *et al.*, 2003). GUS activity was specifically observed in giant cells (not roots) which increased from 60 to 80 per cent at 13-20 dpi depicting a putative role of HSFs in nematode response. Juergensen *et al.* (2003) analysed promoter of the sucrose (Suc) transporter *AtSUC2* gene. Transgenic Arabidopsis plants expressing reporter genes GUS and GFP under the control of *AtSUC2* promoters were inoculated with the beet cyst nematode (*Heterodera schachtii*). GUS and GFP activity was observed as early as 7dpi and at 8 dpi respectively.Thurau *et al.* (2003) studied the expression activity of *Hs1pro-1* gene promoter with reporter gene GUS transformed into susceptible beet roots and Arabidopsis plants. *Hs1pro-1* promoter was up regulated in both species upon infection with beet cyst nematode *Heterodera* *schachtii* and showed a nematode responsive and feeding site-specific GUS-expression. GUS activity was detected at an early stage of nematode infection and was found to be unresponsive in case of wounding.Mazarei *et al.* (2003) analysed *Arabidopsis thaliana* gene *AtPGM* promoter and GUS constructs of transgenic Arabidopsis plants. GUS expression in uninfected plants containing *AtPGM* promoter was localized only to the shoot and root apical meristems. Transgenic plants challenged against *Heterodera schachtii* and *Meloidogyne incognita showed* GUS staining in the nematode feeding structures.Ripoll *et al.* (2003) investigated the potential of mannose binding lectin GNA isolated from *Galanthus nivalis* under the control of the constitutive 35S promoter transformed in *Arabidopsis thaliana,* against the *M. incognita.* Transgenic plants showed a significant decrease (20–50 per cent) in root gall number. Mazarei, (2004) employed the *At17.1* promoter from Arabidopsis fused to GUS reporter gene, and transformed into Arabidopsis plants as well as soybean hairy roots. Strong GUS expression in vascular region upon infection of *H. glycines* was observed. *At17.1* promoter activity was also induced by auxin, jasmonic acid, mannitol and upon dehydration.Mitchum *et al.* (2004) showed that differential expression of tobacco and Arabidopsis driven *Cel1 endo-1, 4-β glucanase* gene promoter in feeding sites induced by cyst (*H. schachtii*) and root knot nematodes (*M. incognita*). The expression of promoter was found to be up-regulated within giant-cell formed by RKN from early to mature giant-cells in both plants. Meutter *et al.* (2005) monitored promoter activity of *ABI3*: : *GUS* in Arabidopsis after *H. schachtii* and *M. incognita* infection. GUS was detected in early infection stages with *H. schachtii*. Alternatively, little or no reporter gene expression was observed upon infection with *M. incognita. ABI3* can be considered to have a putative role during the formation and active growth of the syncytium and demonstrated a marked difference between syncytium and giant cell ontogenesis.Sukno *et al.* (2006), reported an *Arabidopsis thaliana* endoglucanase (*Atcel1*) promoter: GUS construct in transgenic tobacco and Arabidopsis which was found to be activated in giant cells (feeding site). GUS was detected in both shoot and root elongation zones and in the lateral root primordia. Wang *et al.* (2007) isolated the *NtCel7* promoter and analysed its expression at different time points post nematode infection and in response to auxin, gibberellin, ethylene and sucrose in soybean and tomato hairy roots and in Arabidopsis containing the *NtCel7* promoter: GUS construct. Strong GUS expression was observed in root tips and lateral root primordia of uninfected roots. Expression was reported in shoot and root meristems and the vasculature of most organs during plant development. When subjected to hormonal stress, the *NtCel7* promoter was induced only by auxin. Strong GUS activity was observed in both nematode induced feeding sites in transgenic roots of soybean, tomato and Arabidopsis. Grunewald *et al.* (2008) identified a *WRKY* transcription factor expressed during nematode infection. Using both promoter-reporter gene fusions and *in situ* reverse transcription-polymerase chain reaction it was confirmed that *AtWRKY23* expressed during the early stages of feeding site establishment. Wisniewska *et al.* (2013) isolated and evaluated five gene promoters (*CYP97A29, DFR, FLS, NIK* and *PMEI*) against the potato cyst nematode (*G. rostochiensis*). Promoters were fused with the GUS and introduced into tomato and potato genomes via hairy root culture. GUS activity was evident at 7dpi for *NIK, CYP97A29* and at 90 dpi for *NIK.*

6. Model Plant for Nematode-Host Interaction Study

Arabidopsis thaliana is a small plant in the Brassicaceae family that has become the model system of choice for research in plant nematology. Arabidopsis has small seed size of 0.5 mm, short plant height, well established translucent root system which enables study of nematode parasitism easy. The small genome size of 120-megabase is organized into five chromosomes and contains an estimated 20,000 genes, which can be easily transformed. These features make Arabidopsis a model plant for molecular study of nematode-plant interaction (Sijmons *et al.*, 1991).

Conclusion

Plant parasitic nematodes (PPNs) are one of the severe hidden enemies of agricultural as well horticultural crops. Most of the damage contributed by sedentary nematodes namely, root-knot nematodes (*Meloidogyne* spp.), cyst nematodes (*Heterodera* and *Globodera* spp). These tiny organisms have distributed worldwide and lack of appropriate diagnostic tool in fields enables them to multiply in field. This necessitates development of effective management tools to enhance global food security. GM crops are excellent opportunity to control these tiny menaces. For developing transgenics a robust nematode responsive promoter for expressing of gene of interest is immediate need of hour.

REFERENCES

Atkinson, H. J., Urwin, P. E. and McPherson M.J. 2003. Engineering plants for nematode resistance *Annu. Rev. Phytopathol.***41**: 615–39.

Awika, J. 2011. In: Advances in Cereal Science: Implications to Food Processing and Health Promotion; ACS Symposium Series; American Chemical Society: Washington, DC.

Bertioli, D. J., Guimarães, P. M., Jonathon, C. M., Burrows, P. R., and Soraya, C. 2001. Expression of tomato *Cf* genes and their corresponding avirulence genes in transgenic tobacco plants using nematode responsive promoters. *Ann. appl. Biol,***138**: 333-342.

Bertioli, D. J., Smoker, M., Burrows, P. R. 1999. Nematode-responsive activity of the cauliflower mosaic virus 35S promoter and its sub domains. *Mol Plant Microbe Interact,***12**: 189-196.

Blaxter, M. L. 2000. Nematoda: genes, genomes and the evolution of parasitism. *Adv. Parasitol.* **54:** 101-195.

Cobb, N. A. 1918. Estimating the nema population of the soil, with special reference to the sugar-beet and root-gall nemas, *Heterodera schachtii* (Schmidt) and *Heterodera radicicola* (Greef) Muller, and with a description of *Tylencholaimus aequalis* n. sp. USDA. *Agr. Tech. Circ.* **1:** 47.

Cook, R., and K. Evans. 1987. Resistance and tolerance. P. 179-231 in R. H. Brown and B. R. Kerry (ed.). Principles and practice of nematode control in crops. Marrickville, NSW, Australia: Academic Press Book.

Coutu, C., Brandle, J., Brown, D., Brown, K., Miki, B., Simmonds, J., Hegedus D. D. 2007. A modular binary vector series suited for monocot and dicot plant transformation. *Transgenic Res.* **16**: 771-781.

Decraemer, W., Hunt, D. J. 2006. Structure and classification. In: Perry R.N, Moens, M (eds) Plant nematology. CABI Publishing, Wallingford, pp 3–32.

Dinh, P. T. Y., Zhang, L., Brown, C. R., Elling, A. A. 2014. Plant-mediated RNA interference of effector gene *Mc16D10L* confers resistance against *Meloidogyne chitwoodi* in diverse genetic backgrounds of potato and reduces pathogenicity of nematode offspring. *Nematology*. **16**: 669-682.

Dutta, T. K., Banakar, P and Rao, U 2015. The status of RNAi-based transgenic research in plant nematology. *Front.Microbiol.***5**: 760.

Ehsanpour, A.A., Jones, M.G.K. 1996. Glucuronidase expression in transgenic tobacco roots with a *Parasponia* on infection with *Meloidogyne* javanica. *J. Nematol*. **28**: 407-413.

Elling, A. A. 2013. Major emerging problems with minor *Meloidogyne* species. *Phytopathology*. **103**: 1092-1102.

Escobar, C. D., Meutter, J., Aristizábal, F.A., Sanz-Alféez, S., Barthels, N., Eycken, W., Seurinck, J., Montagu, M., Gheysen, G and Fenoll, C. 1999. Isolation of the *LEMMI9* gene and promoter analysis during a compatible plant–nematode interaction. *Molecular Plant-Microbe Interactions*. **12**: 440–449.

Escobar, C., Barcala, M., Portillo, M., Almoguera, C., Jordano, J., Fenoll, C. 2003. Induction of the *Hahsp17.7G4*promoter by root-knot nematodes: Involvement of heat-shock elements in promoter activity in giant cells. *Molecular Plant-Microbe Interactions*. **16:** 1062-1068.

Ferris,T., Bongers B, R.G.M. de Goedec. 2001. A framework for soil food web diagnostics: extension of the nematode faunal analysis concept. *Applied Soil Ecology*. **18:** 13-29.

Gaur, H. S. and Singh, I. 1977. Pigeon-pea cyst nematode, *Heterodera cajani*, associated with the moong crop in the Punjab State. *Journal of Research*, Punjab Agricultural University, **14(4)**: 509.

Gelvin, B. S. 2003. Agrobacterium-Mediated Plant Transformation: the Biology behind the "Gene-Jockeying" Tool. *Microbiology and Molecular Biology Reviews*. **67(1)**: 16-37.

Golden, A.M. and Birchfield, W. (1965). *Meloidogyne graminicola* (Heteroderidae) a new pecies of root-knot nematode from grass. *Proc. Helmintholog. Soc. Wash.* **32**: 228-231.

Goverse, A., Biesheuvel, J., Wijers, G. J., Gommers, F. J., Bakker, J., Schots, A., Helder, J. 1998. *In planta* monitoring of the activity of two constitutive promoters, CaMV 35S and TR2 in developing feeding cells induced by *Globodera rostochiensis* using green fluorescent protein in combination with confocal laser scanning microscopy. *Physiol Mol Plant Pathol*. **52**: 275-284.

Grunewald, W., Karimi, M. Wieczorek, K., Van de Cappelle, E., Wischnitzki, E., Grundler, F., Inze, D., Beeckman, T., Gheysen, G. 2008.A role for *AtWRKY23* in feeding site establishment of plant-parasitic nematodes. *Plant Physiol.***148**: 358-368.

Hansen, E., Harper, G., Mcpherson, M.J., Atkinson, H.J. 1996. Differential expression patterns of the wound-inducible transgene *wun-1-uidA* in potato roots following infection with either cyst or root-knot nematodes. *Physiol Mol Plant Pathol.* **48**: 161-170.

Hepher, A. and Atkinson, H.J. (1992) Nematode control with Proteinase inhibitors. European patent publication number 0 502 730 A1.

Huang, G., R. Dong, R. Allen, E. L. Davis, T. J. Baum, and R.S. Hussey. 2006. Engineering broad root-knot resistance in transgenic plants by RNAi silencing of a conserved and essential root-knot nematode parasitism gene. *Proc. Natl. Acad. Sci. USA*. **103**: 14302-14306.

Hutangura, P., Mathesius, U., Jones, M. G. K., Rolfe, B.G. 1999. Auxin induction is a trigger for root gall formation caused by root-knot nematodes in white clover and is associated with the activation of the flavonoid pathway. *Aust J Plant Physiol.* **26**: 221-231.

Jain, R. K., Mathur, K. N., and Singh, R. V. 2007. Estimation of losses due to plant parasitic nematodes on different crops in India. *Indian Journal of Nematology.* **37(2):** 219-221.

Jefferson, R. A., Kavanagh, T. A. and Bevan, M. W. 1987. GUS fusions: *b-glucuronidase* as a sensitive and versatile gene fusion marker in higher plants. *EMBO J.* **6**: 3901-3907.

Jones JT, Haegeman A, Danchin EG, Gaur HS, Helder J, Jones MG, Kikuchi T, Manzanilla-López R, Palomares-Rius JE, Wesemael WM, Perry RN. 2013. Top 10 plant-parasitic nematodes in molecular plant pathology. *Molecular Plant Pathology*, **14(9)**: 946-61.

Juergensen, K., Scholz-Starke, J., Sauer, N., Hess, P., van Bel, A. J. E., and Grundler, F. M. W. 2003. The companion cell-specific Arabidopsis disaccharide carrier *AtSUC2* is expressed in nematode-induced syncytia. *Plant Physiology*. **131**: 61-69.

Karimi, M., Manes, O., Montagu, M. V. and Gheysen, G. 2002. Activation of a Pollenin Promoter upon Nematode Infection. *Journal of Nematology*. **34(2)**: 75-79.

Linford, M. B, Oliveira, J. M. 1940. *Rotylenchulus reniformis*, nov.gen. n. sp., a nematode parasite of roots. *Proceeding of the Helminthological Society of Washington*. **7**: 35-42.

MacGowan, J.B. and Longdon, K.R. Host of rice root knot nematode *Meloidogyne graminicola*, Nematology circular, Gainesville No. 172.

Mazarei, M., Kristen, A., Lennon, David, P., Puthoff, Steven, R., Rodermel and Thomas, J. B. 2004. Homologous soybean and Arabidopsis genes share responsiveness to cyst nematode infection. *Molecular Plant Pathology*. **5**: 409-423.

Meutter, L. D., Robertson, L., Parcy, F., Mena, M., Gheysen, C. F. 2005. Differential activation of *ABI3* and *LEA* genes upon plant parasitic nematode infection. *Molecular Plant Pathology*. **6(3)**: 321-325.

Mitchum, M., Sukno, S., Shani, Z., Shoseyov, O., Davis, E. L. 2004. The promoter of the *Arabidopsis thaliana cel1* Endo-1, 4-β-glucanase gene is differentially expressed in plant feeding cells induced by root-knot and cyst nematodes. *Molecular Plant Pathology*. **25**: 175-181.

Mitchum, M. G *et al.*, 2012. Role of Nematode Peptides and Other Small Molecules in Plant Parasitism. *Annual review of phytopathology*. **50**: 175-95.

Moller, S. G., Urwin, P. E., Atkinson, H. J., and McPherson, M. J. 1998. Nematode-induced expression of *atao1*, a gene encoding an extracellular diamine oxidase associated with developing vascular tissue. *Physiological and Molecular Plant Pathology*. **53**: 73-79.

Mulk, M. M. 1976. *Meloidogyne graminicola*. C.I.H. Descriptions of Plant-parasitic Nematodes. **(6)**: 87: 4.

Ohl, S. A., van der Lee, F.M., Sijmons, P. C. 1997. Anti-feeding structure approaches to nematode resistance. Regulation of gene expression in feeding sites. In: Fenoll C, Grundler FMW, Ohl SA, editors. Cellular and Molecular Aspects of Plant-Nematode Interactions. Dordrecht: Kluwer Academic Publishers; 1997. pp. 250–261.

Opperman, C.H. and Conkling, M.A. (1996) Analysis of transgenic root knot nematode resistant tobacco. *Abstract. Third International Nematology congress, Guadeloupe.*

Opperman, C. H., Taylor, C. G. 1994. Conkling M. A. Root-knot nematode directed expression of a plant root-specific gene. *Science*. **263**: 221-223.

Papolu, P. K., Gantasala, N. P., Kamaraju, D., Banakar, P., Sreevathsa, R., Rao, U. 2013. Utility of host delivered RNAi of two FMRF amide like peptides, *flp-14* and *flp-18*, for the management of root-knot nematode, *Meloidogyne incognita*. PLoS ONE 8: e80603.

Prasad, J. S. and Varaprasad, K. S.(2001). *Ufra* nematode, *Ditylenchus angustus* is seed borne. *Crop Protec*. **21**(1): 75-76.

Puzio, P.S, Lausen, J., Heinen, P., Grundler, F. M. W. 2000. Promoter analysis of *pyk20*, a gene from *Arabidopsis thaliana*. *Plant Sci*. **157**: 245-255.

Ripoll, C., Favery, B., Lecomte, P., Damme, E.V. Peumans, W., Abad, P., Lise Jouanin. 2003. Evaluation of the ability of lectin from snowdrop (*Galanthus nivalis*) to protect plants against root-knot nematodes. *Plant Science*. **164**: 517-523.

Roberts, P.A. (1992) current status of the availability, development and use of host plant resistance to nematodes. *Journal of Nematology*, **24:** 213-227.

Serenella, S., Orit, S., Jamie, M., Galit, T., Ziv, S., and Oded, S. 2006. Expression and Regulation of the *Arabidopsis thaliana Cel1* Endo 1, 4 Glucanase Gene during Compatible Plant-Nematode Interactions. *Journal of Nematology*. **38(3)**: 354-361.

Sharma, S. B., Reddy, B. M. R., Krishnappa, K. 1992. Incidence and importance of plant-parasitic nematodes on pigeon pea and groundnut in Karnataka State in southern India. *Afro-Asian Journal of Nematology*. **2(1-2)**: 21-26.

Sharma, S. B., Nene, Y. L, Reddy, M. V., McDonald, D. 1993. Effect of *Heterodera cajani* on biomass and grain yield of pigeon pea on vertisol in pot and field experiments. *Plant Pathology*. **42(2)**: 163-167.

Sijmons, P. C., Grundler, F. M. W., Vonmende, N., Burrows, P. R., Wyss, U. 1991. *Arabidopsis thaliana* as a new model host for plant-parasitic nematodes. *Plant J*.**1**: 245-254.

Thurau, T., Kifle, S., Jung, C., Cai, D. 2003. The promoter of the nematode resistance gene *Hs1pro-1* activates a nematode responsive and feeding site specific gene expression in sugar beet (*Beta vulgaris* L.) and *Arabidopsis thaliana*. *Plant Mol Biol*. **52**: 643-660.

Urwin, P. E., McPherson, M. J. and Atkinson, H.J. 1998. Enhanced transgenic plant resistance to nematodes by dual proteinase inhibitor constructs. *Planta* 204. **(4)**: 472-479.

Urwin, P. E., Moller, S. G., Lilley, C. J., McPherson, M. J., Atkinson, H.J. 1997. Continual green-fluorescent protein monitoring of cauliflower mosaic virus 35S promoter activity in nematode-induced feeding cells in *Arabidopsis thaliana*. *MPMI*. **10(3)**: 394-400.

Urwin, P.E., Mcpherson, M.J. and Atkinson, H.J. (1998) Enhanced transgenic plant resistance to nematodes by dual proteinase inhibitor constructs. *Planta*,**204:** 474-479.

Wang, X., Replogle, A., Davis, E. L., Mitchum, M. G. 2007. The tobacco *Cel7* gene promoter is auxin-responsive and locally induced in nematode feeding sites of heterologous plants. *Mol Plant Pathol*. **8**: 423-436.

Wang, D.Y.-C., S. Kumar, and B.S. Hedges. 1999. Divergence time estimates for the early history of animal phyla and the origin of plants, animals and fungi. *Proceedings of the Royal Society of London*. **B 266**: 163-171.

Weibelzahl-Fulton, E., Dickson, D. W. and Whitty, E. B. 1996. Suppression of *Meloidogyne incognita* and *M. javanica*.

Wis´niewska, J., Da²browska-Bronk, K., Szafran´ski, S., Fudali, M., S´wie²cicka, M., Czarny, A., Wilkowska, K., Morgiewicz, J., Matusiak, M., Sobczak, M., Filipecki. Analysis of tomato gene promoters activated in syncytia induced in tomato and potato hairy roots by *Globodera rostochiensis*. *Transgenic Res*. 2013. **22**: 557-569.

Yadav, B. C., Veluthambi, K., and Subramaniam, K. 2006. Host generated double stranded RNA induces RNAi in plant- parasitic nematodes and protects the host from infection. *Molecular and Biochemical Parasitology*, **148:** 219-222.

Zuckerman, B. M., and Jansson, H. 1984. Nematode Chemotaxis and Possible Mechanisms of Host/Prey Recognition. *Annual Review of Phytopathology*. **22**: 95-113.

Conservation Agriculture Technologies *Pages* **183–190**
Editor: **Rahul Kumar, Abhinandan Singh and Pankaj Kumar Ojha**
Published by: **BIOTECH BOOKS, NEW DELHI**

16 Insect Pheromones and their Role in Insect Pest Management

Ram Kumar and I. Yimjenjang Longkumer

Department of Entomology, Dr. Rajendra Prasad Central Agricultural University, Pusa, Samastipur – 848 125, Bihar

Insect pheromones are one of the most important tools in integrated pest management as it works very efficiently in detection and monitoring, mass trapping and mating disruption of insect's pests in eco-friendly manner. The pheromones are required in very minute quantity to attract and kill a large number of insect pests and hence they are economical. Use of pheromones alone is not much more effective as when they are used with insecticides. Pheromone technology has potential to add value in long-term pest management of many economically important pests. Inclusion of pheromone formulations in integrated pest management does not cause any deleterious or harmful effect on human and their interest. Pheromone technology is cost effective and applicable in area wide pest management approach.

Introduction

The chemical compounds, which modify the behaviour of perceiving organisms, are called as 'semiochemicals'. These chemicals may facilitate both intra and inter-specific communication in an organism. On the basis of intra and inter-specific communication semiochemicals are classified into allelochemicals and pheromones.

Allelochemicals

The term allelochemical was proposed by R.H. Whittaker (1970), for certain chemical compounds, which trigger a behavioural change in inter-specific species.

Allelochemical are further divided into:

1. Kairomone: adaptively favourable to receiver but not to emitter.
   It may be attractants, arrestants or exitants.
2. Allomone: adaptively beneficial to emmiter but not receiver.
   It may be repellent, determnent, antibiotics or antixenotics.
3. Synomone: adaptively favourable to both emitter and receiver.
4. Apneumones: the chemical compound released by non-living a non-living material which triggers a behavioural change that is favourable to a receiver but detrimental to another found on the living substance.
5. Antimones: a chemical released by an individual, when contact with an individual of another species, causes a physiological or behavioural modifications which is harmful to both emitter and receiver.

Pheromones

Karlson and Luscher (1959) coined the term "pheromone" to represent the chemical compounds that enable insects for intra-specific communication. It is also referred as "ecto-hormone". The term pheromone was derived from the greek "pherein" (to carry) and "horman" (to excite, stimulate). Pheromones are a class of semiochemicals or info-chemicals that insects and other animals elicit, to communicate with other individuals of the same species and causes specific response in the receiving organism. They are active (*e.g.* attractive) in extremely low doses (one millionth of an ounce) and are used to bait traps or confuse a mating population of insects.

Types of Pheromones

Insect pheromones are classified as follows:

Primer

These pheromones trigger a change of developmental events (in which they differ from all the other pheromones, which trigger a change in behaviour). A primer pheromone has slower onset and longer effect.

Releaser

Releaser pheromones are pheromones that cause an alteration in the behaviour of the recipient. For example, some organisms use powerful attractant molecules to attract mates from a distance of two miles or more. In general, this type of pheromone elicits a rapid response, but is quickly degraded.

Sex Pheromones

Sex pheromones are chemical compound released by an organism to attract an individual of the opposite sex, encourage them to mate with con-specifics or perform some other function closely related with sexual reproduction. It is usually reffered to as "sex attractant" or "sex lure". The first described chemical was bombykol, the silkworm moth's sex pheromone, which is a complex alcohol, trans-

10-cis-12 hexadecadienol, discovered in 1959 by Karlson and Butenandt. Generally male respond to odorous chemical released by female, but in some species males are known to produce sex pheromone also, which attracts the female. Insects in which male produce sex pheromones are; cotton boll weevil, *Anthonomus grandis*, cabbage looper, *Trichoplusia ni* and mediterranian fruit fly, *Ceratitis capitata*. The sex pheromones are most powerful chemical among the chemical attractants. Sex pheromones specifically focus on indicating females for breeding, attracting the opposite sex, and conveying information on species, age, sex and genotype. The effectiveness of the sex pheromone depends upon the threshold concentration for male stimulation and release rate from the female. Different species use a wide variety of chemical substances to send sexual signals. The sex pheromone isolated and identified in many insects are; pink bollworm, *Pectinophora gossiypiella*, 10-propyl-trans-5,9-tridecadienyl acetate; gypsy moth, *Porthetria dispar*, 10-acetoxy-cis-7-hexadecenol; cabbage looper, *Trichoplusia ni*, cis-7-dodecenyl acetate; honeybee, *Apis mellifera*, 9-keto-trans-2-decenoic acid.

Alarm Pheromones

A pheromone released by an animal, especially a social insect, in response to attack or injury, typically triggering a response of aggression or flight in others of the same species or colony. A large literature addresses chemical identification of alarm pheromones and their impact on the behaviour of nearby individuals. In order to conclude that particular compounds acts as an alarm pheromone, it is generally considered necessary to demonstrate that (i) the chemical(s) is released exclusively under exposure to hazard (*e.g.* predator attack), (ii) the signal is perceived by con-specifics, and (iii) it induces in the receiving individuals behavioural reactions similar to that induced when directly exposed to the same danger (Wyatt, 2003). Alarm pheromones are reported in insect order *viz.* Homoptera, Isoptera and Hymenoptera.

E.g.: - Ants (released through anal, mandibular, dufours and poison glands). When an ant is disturbed, it releases a pheromone that can be detected by other ants several centimeters away. They are attracted by low concentrations of the pheromone and begin to move toward the region of increasing concentration. As they get nearer to their disturbed nest mate, their response changes to one of alarm. The higher concentration causes them to run about as they work to remedy the disturbance. Unless additional amounts of the alarm pheromone are released, it soon dissipates. This ensures that once the emergency is over, the ants return quietly to their former occupations. In some insects alarm pheromones are reported and released by organs like; cephalic glands (termites), sting apparatus and mandibular glands (worker bees) and aphids (cornicals).

Aggregation Pheromones

Aggregation pheromones function in mate selection, overcoming host resistance by mass attack, and defense against predators. Male-produced sex attractants often are referred to as aggregation pheromones because they typically result in the arrival of both sexes at a calling site (Matthews and Matthews, 1978; Jutsum and Gordon, 1989; Landolt, 1997). The aggregation pheromones draw more than two individuals together. Virgin bark beetles females (scolytidae) release pheromones that attract both sexes to mating and oviposition sites thus aggregating large number of beetles to recently damaged timber (Lanier *et al.*, 1972). A rather small percentage of sex attractants is produced by males (Jutsun and Gordon, 1989; McCaffery and Wilson, 1990; Landolt, 1997). Most of the insects that use male-produced sex attractant are found within the Coleoptera and according to Landolt (1997), up to 1989 males pheromones attractive to females had been identified for 54 species. Forty of these species are coleopterans, 9 are dipterans, 2 are hemipterans, and 3 are lepidopterans. In addition, he found that since 1988 another 40 males pheromones attractive to females species were identified; 18 coleopterans, 12 dipterans, 7 hemipterans, 2 lepidopterans and 1 dictyopteran. The production of aggregation pheromones by several weevils, including *Anthonomus grandis*, three *Sitophilus* spp., and *Sitona lineatus* is well documented (Hardee *et al.*, 1969; Tumlinson *et al.*, 1969; Phillips *et al.*, 1981; Blight *et al.*, 1987). According to Jutsum and Gordonn (1989), the management of these and other major pest insects using aggregation pheromones has been examined. Overall the use of pheromones has a number of potential advantages. The compounds are naturally occurring, generally non-toxic and should not pollute the environment. Pheromones are insect-specific and their safety to beneficial, makes them ideal components of integrated pest management systems.

Trail-Marking Pheromones

These pheromones are mainly reported in social insects such as ants, termites *etc.* The certain ants, as they return to the nest with food, lay down a trail pheromone. This trail attracts and guides other ants to the nest. It is continually renewed as long as the food holds out. When the supply begins to dwindle, trail making ceases.

The trail pheromone evaporates quickly *i.e.* have low persistence, so other ants stop coming to the site and are not confused by old trails when food is found elsewhere. A stick treated with the trail pheromone of an ant can be used to make an artificial trail which is followed closely by other ants emerging from their nest. Some other examples are; ants, (*formica rufa*) formic acid, hexanonic and heptanoic acid; caproic acid (*Zootermopsis* sp.).

Epideictic Pheromones

Epideictic pheromones are also known as spacing pheromone. These are the chemicals which stimulate dispersal of the individual from a source when an optimal density has been reached. These pheromones limit the intraspecific competition, in fruit flies after oviposition fly produce a chemical which inhibit the oviposition of the same fruit by another fruit fly. Bark beetle produces some chemical which reduces the further arrival of individuals on host trees.

Host marking Pheromones secreted by many fruit flies (Family: tephritidae) provide excellent examples of such pheromones. Certain pheromones released by bark beetles reduce further arrivals of individuals on host trees.

Pheromone Exploitation

The insect pheromones, has a wide application in integrated pest management system, as a tool of great importance for insect control. At present these chemicals are used as much safer weapons to control the pest. These are eco-friendly and are more species- specific. Among all the sex pheromone is the most important for integrated pest management. It can be practically employed for monitoring, survey trapping, mass trapping and mating disruption in integrated pest management strategy. The pheromones are mainly used in the integrated pest management as follows:

1. Detection and Monitoring

The most important application of pheromone is, to evaluate the pest population size, to determine pest distribution pattern and outbreak detection. Traps baited with pheromone lures, are used to estimate the insects population and to detect the insect infestation in a new area. The monitoring function is the keystone of integrated pest management. Monitoring is used extensively in urban pest control of cockroaches, in the management of stored grain pests in warehouses or distribution centres, and to track the nationwide spread of certain major pests such as the gypsy moth. The most widespread use of pheromones has been for monitoring endemic pest species' adult populations. Comprehensive apple IPM programs that were initiated in New York State and Michigan during the early 1970s were based on good, species-specific monitoring traps for the complex of tortricid moth pests that cause direct and indirect damage to fruit. Monitoring of leafroller pests coupled with computer assisted degree-daymodels (Riedl and Croft, 1974; Riedl *et al.*, 1976) allowed sprays to be timed for optimum efficacy against eggs and first instars on such key pests as the codling moth (*Cydia pomonella*) and oriental fruit moth (*Grapholita molesta*). Pheromone traps have played a large role in detecting influxes of adult pest species from one region to another and also even from non-crop areas into crop areas. Survey programs involving grids of pheromone traps are used so routinely to report and track the yearly arrival of migrating adult populations of insects such as the black cutworm (*Agrotis ipsilon*) in the Midwest (Showers *et al.*, 1989) or the spread of expanding populations such as the gypsy moth (*Lymantria dispar*) (Elkinton and Card´e, 1981) that it has become an essential part of our arsenal of tools for tracking and ameliorating pest-movement-related threats.

2. Mass Trapping

Second major use of pheromones is, mass trapping of insects to remove large numbers of insects from breeding and feeding population. Use of traps on large scale in the field ensures that large no. of male individuals will be trapped which significantly reduces the chance of mating and fertilization thus reduces production of new generation of insect pests. Individuals will be trapped, which significantly reduces chance of mating and fertilization and thus production of new generation of insect's pests. Massive reductions in the population density of pests ultimately help to protect resources such as food or fibre for human use. Mass trapping has been explored with pine bark beetles and has resulted in millions of insects attracted specifically into traps and away from trees. Mass trapping has been used successfully against the codling moth, a serious pest of apples and pears. This technique is mostly restricted to protection of crops where the insecticides are not effective and environmentally acceptable. Likewise food lures used in combination with pheromones may offer a way of enhancing the effectiveness of the mass-trapping system for stored-product pests (Cox, 2004). Recent studies have investigated the potential of pheromone based mass-trapping methods to control indoor population of *Ephestia kuehniella* (Anderbrant *et al.*, 2007; Süss *et al.*, 1996). (Trematerra and Battaini, 1987) demonstrated that integrated control of *E. kuehniella* can be achieved by mass-trapping.

3. Mating Disruption

Pheromones are used extensively in mating disruption of insect pests. This has been most effectively used for Lepidopteran pests. In this scenario, synthetic pheromone is dispersed into crop fields and the false odour plumes attract males away from females that are waiting to mate. This causes a reduction of mating, and thus reduces the population density of the pests. In some cases, the effect has been so great that the pests have been locally eradicated. Many such pheromones are available commercially and their use is considered as a "bio-rational" approach to insect control. Mating disruption, due to the specificity of the sex pheromone of the insect species, has the benefit of only affecting the males of that species, while leaving other non target species unaffected. This allows for much targeted pest management, promoting the suppression of a single pest species, leaving the populations of beneficial insects (pollinators and natural enemies) intact. Mating disruption, like most pest management strategies, is a useful technique, but should not be considered a stand-alone treatment program (Foster and Harris, 1997) because it targets only a single species in plant production systems that usually have several pests of concern. The use of synthetic pheromone "Gossyplure" with regard to cotton pink bollworm, *Pectinophora gossypiella* has been successfully demonstrated in India. At present, in India synthetic pheromone of *H. armigera, Earias insulana, E. vitella, Pectinophora gossypiella, Spodoptera litura, Plutella xyllostella, Leucinodes orbonalis, Orycetes rhinoceros* and *Rhynchophorus ferrugineus* are commercially available.

Conclusion

At present pheromones are used as much safer weapons to control the agriculturally important insects' pests. It can be practically employed for monitoring, survey trapping, mass trapping and mating disruption in integrated pest management strategy. The most widespread use of pheromones has been for monitoring endemic pest species' adult populations. The efforts are also given to make regarding ways to determine empirically the modes of action of mating disruption formulations, and the acceptance of the mating disruption technique by growers and government agencies has continued to grow in recent years. The pheromone lure used for insect control, at present, is species specific which is not feasible at farmers level thus, there is need to develop cocktail approach *i.e.* mixing together several pheromones in single lure.

REFERENCES

Allison, Jeremy, D and Carde, R.T. (2016). *Pheromone Communication in Moths: Evolution, Behavior, and Application*. University of California Press. pp. 265–271.

Anderbrant, O., Ryne, C., Olsson, P.O.C., Jirle, E., Johnson, K and Löfstedt, C. (2007). Pheromones and kairomones for detection and control of indoor pyrali moths. IOBC/wprs Bulletin. **30**: 73-77.

Benjamin, H.S., Leal, L.N., Walter, S and Jon, C. (2000). "Sexual attraction in the silkworm moth: structure of the pheromone-binding-protein-bombykol complex". *Chemistry and Biology*. **7**(2): 143-151.

Blight, M.M and Wadhams L.J. (1987). Male-produced aggregation pheromone in pea and bean weevil, *Sitona lineatus* (L.). *Journal of Economic Entomology*. **13**: 4.

Cox, P.D. (2004). Potential for using semio-chemicals to protect stored products from insect infestation. *Journal of Stored Products Research*. **40**: 1-25.

Elkinton, J.S and Carde, R.T. (1981). The use of pheromone traps to monitor distribution and population trends of the gypsy moth. In *Management of insect pests with semiochemicals*, ed. E.R Mitchell, pp. 41-55. New York: Plenum Press.

Foster, S.P and Harris, M.O. (1997). Behavioral manipulation methods for insect pest-management. *Annual Review of Entomology*. **42**: 123–146.

Hardee, D.D., Cross, W.H and Mitchel, E.B. (1969). Male boll weevil are more attractive than cotton plants to boll weevil. *Journal of Economic Entomology*. **62**: 165.

Jutsum, A.R and Gordon, R.F.S. (1989). Pheromones: importance to insects and role in pest management. In Insect pheromones in plant protection. Ed. A.R. Jutsum and R.F.S. Gordon. John Wilen and Sons Ltd. pp. 1-13.

Karlson, P and Butenandt, A. (1959). *Ann. Rev. Entomol.* **4**: 39.

Landolt, J.P. (1997). Sex attractant and aggregation pheromones of male phytophagous insects. *American Entomologist*. **43**: 1.

Lanier, G.N., Birch, M.C., Schmitz, R.F and Furmiss, M.M. (1972). Pheromones of IPS pini (Coleoptera: scolytidae) variation in response among three population. *Can. Ent.* **104**: 1917-1923.

Matthews, R.W and Matthews. (1979). Insect behaviour. New York: Wiley C.

McCaffery, A.R and Wilson, I.D. (1990). Chromatography and isolation of insect hormones and pheromones. Ed. Plenum press. New York and London.

Phillips, J.K and Burkholder, W.E. (1981). Evidence for a male-produced aggregation pheromones in the rice weevil. *Journal of Economic Entomology*. **74**: 539-542.

Riedl, H and Croft, B.A. (1974). A study of pheromone trap catches in relation to codlingmoth (Lepidoptera: olethreutidae) damage. *Canadian Entomologist*, **112**, 655-663.

Riedl, H., Croft, B.A and Howitt, A.G. (1976). Forecasting codling moth phenology based on pheromone trap catches and physiological-time models. *Canadian Entomologist*, **108**, 449-460.

Sarkar, S and Patra, S. (2016). Pheromone: A key tool in integrated pest management. *Rashtriya Krishi.*, **11**(1): 63-64.

Showers, W.B., Whitford, F., Smelser, R.B. (1989b). Direct evidence for meteorologically driven longrange dispersals of an economically important moth. *Ecology*, **70**: 987-992.

Süss, L., Locatelli, D.P. and Marrone, R. (1996). Possibilities and limits of mass trapping and mating disruption techniques in the control of *Ephestia kuehniella* Zeller (Lepidoptera Phycitidae). *Bollettino di Zoologia agrarie di Bachicoltura*. **28**: 77-89.

Trematerra, P and Battaini, F. (1987). Control of *Ephestia kuehniella* Zeller by mass-trapping. *Journal of Applied Entomology*. **104**: 336–340.

Tumlinson, J.H., Hardee, D.D., Gueldner, R.C., Tompson, A.C., Hedin, A.P. and Minyard, J.P. (1969). Sex pheromones produced by male boll weevils. *Isolation, identification and synthesis science*. **166**: 1010-1012.

Wyatt, T.D. (2003). Pheromones and animal behaviour. The press syndicate of the University of Cambridge, Oxford.

Conservation Agriculture Technologies *Pages* **191–209**
Editor: **Rahul Kumar, Abhinandan Singh and Pankaj Kumar Ojha**
Published by: **BIOTECH BOOKS, NEW DELHI**

17 Resource Conserving Technologies: Way Forward for Sustainable Management of Natural Resources and Climate Change Mitigation

***Suman Sen*[1] *and Vijay Kant Singh*[2]**

***[1]Ph.D. Scholar, Division of Agronomy,
ICAR-Indian Agricultural Research Institute, New Delhi – 110 012
[2]Ph.D. Scholar, Department of Soil Science, Govind Ballabh Pant University of Agriculture and Technology, Pantnagar – 263 145, Uttarakhand***

The Green Revolution (GR) during late 1960s, helped the nation in achieving food sufficiency. Various GR technologies (incorporating the use of high-yielding varieties, fertilizers, pesticides and irrigation water) were used to a greater extent for enhancing the production. Continuous intensive use of GR technologies has led to the emergence of many second generation problems like, rising or lowering of water-table, deteriorating soil and environment quality, nutrient imbalance, resurgence of pest and diseases, stagnation of yield and lower marginal returns, besides the concern of continuous climate change. Moreover, over-exploitation on natural resources has resulted in tremendous pressure on natural resource bases that is reaching the level of ecological crises. Therefore, the issue of sustainability is the major challenge of present day agriculture. Further, escalating cost of agricultural inputs necessitates the effective utilization of resources in agriculture. It is, therefore, the need of the time to adopt resource conserving technologies (RCTs) such as, zero or reduced tillage systems with direct seeding and bed planting, management of crop residue and nutrients, and crop diversification for efficient management of soil, water and nutrient. The RCTs, an exponent of conservation agriculture (CA), have the potential to improve the production levels as well as quality, soil biodiversity and use-efficiency of natural resources like water, air, fossil fuel and soil. By adopting the CA-based RCTs, the need of present generation can be fulfilled without endangering the resource bases for the generations to come. These technologies, over the period can improve the sustainability of agricultural production system and improve the natural resource bases. These, further improve the quality of environment by mitigating greenhouse gas emissions and reducing residue burning and help in adapting to climate change. However, these new technologies are site-specific and there is a need for wider scale promotion and assessment of these technologies across various production systems, agro-ecologies, cropping seasons and farm conditions to have significant adoption.

Introduction

Indian agriculture has the challenge of providing the national as well as household food and nutritional security from the limited and continuously shrinking agricultural land. During the past fifty years, the human population of India has doubled to 1.21 billion and likely to increase to more than 1.6 billion before near-stabilization by 2030 with subsequent more per capita use of natural resources. The food demand is expected to rise up to ~400 million tonnes by the year 2050 (Anonymous, 2015). The prospect of horizontal expansion of area under farming is limited as agricultural land is continuously shrinking due to continuous encroachment of good land by expanding urbanization and industrialization. It is, therefore, imperative to enhance the productivity with efficient and judicious use of natural resources. The green revolution during late 1960s, helped the nation to become self-sufficient in terms of foodgrain production but it compromised with environmental quality. To achieve this goal, various critical inputs like new varieties, pesticides, fertilizers and irrigation water were used to a greater extent for enhancing the production. As a result foodgrain production has increased more than 5-fold during the past 50 years but resource-use efficiency was hardly taken into consideration. Over exploitation and injudicious use of the resource base has resulted in unprecedented pressure on natural resources that is reaching the level of ecological crises. All these have led to the emergence of many second generation problems like, rising or lowering of water-table, soil degradation, nutrient imbalance, salinity, resurgence of pest and diseases, declining natural resource bases, decline in soil bio-diversity, pollution of water and air with toxic substances, stagnation of yield and farmers' income besides the concern of continuous climate change (Das *et al.*, 2014).

Therefore, the major challenge of present day agriculture is to achieve improved productivity with less resources use without deteriorating soil and environment quality, thereby ensuring long-term sustainability with a suitable balance in agro-ecosystem (Kumar and Ladha, 2011). The looming crises pertaining to ecological imbalance in agriculture sector drive the search for appropriate alternative resource conserving technologies (RCTs) to keep pace with the ever increasing human population demand. Resource conserving technologies (RCTs) have the potential of sustaining the production levels as well as quality and enhancing the input-use efficiency. By adopting RCTs, the need of the present generation can be fulfilled without endangering the resource bases for the generations to come. This, over the period, can improve the quality of environment, as well. Therefore, during last two decades more emphasis has been given worldwide on development of RCTs which would lead to a drastic reduction in production costs and enhance farmers' income (Das *et al.*, 2014).

Resource Conserving Technologies vis-à-vis Conservation Agriculture: Concept and Definition

Resource conservation technologies in agriculture are same as the name suggest, it is the conservation or saving of the resources or inputs without compromising with the output/yield. In simpler words, RCTs are the practices that enhance the output per unit of input. Therefore, resource conserving technologies can be defined as the tools and techniques which have the potential to sustain the crop production level vis-à-vis quality without endangering the natural resource bases for future generation with more focused approach on improving input/resource-use efficiency and conserving resource bases.

The conservational approach in the agriculture is the major demand in today's scenario. Sustainability of agriculture evolved the concept of conservation agriculture (CA), globally. At present, CA is practiced on around 156.9 M ha worldwide which is about 11 per cent of the world's arable land (Table 17.1).

Table 17.1: Adoption of Conservation Agriculture

Country	*Area (M ha)*	*Per cent of Global Area*
USA	35.61	22.68
Brazil	31.81	20.26
Argentina	29.18	18.59
Canada	18.31	11.67
Australia	17.70	11.27
China	6.67	4.25
Russian federation	4.50	2.87
Paraguay	3.00	1.91
Kazakhstan	2.00	1.27
India	1.50	0.96
Others	6.71	4.27
Total	**156.99**	**100**

Source: FAO (2017).

Conservation agriculture as defined by Food and Agriculture Organization (http: //www.fao.org/ag/ca) is a resource-saving agricultural production system that enhances natural and biological processes above and below ground. Conservation agriculture systems utilize soils for the production of crops with the aim of reducing excessive mixing of the soil and maintaining crop residues on the soil surface in order to minimize damage to the environment (Derpsch, 2001). So, CA is a resource-saving agricultural production system that aims to achieve production intensification and high yields while improving the natural resource bases, including water, air, fossil fuels, soils, inputs and human resources (Abrol and Sangar, 2006).

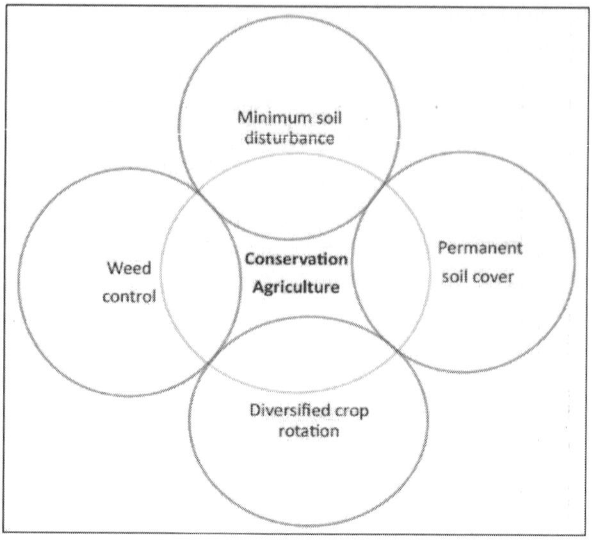

Figure 17.1: Principles of Conservation Agriculture (*Source*: Farooq and Siddique, 2015).

Conservation agriculture is characterized by four interrelated principles (Figure 17.1) along with other good production practices of plant nutrition and pest management, *viz.* (i) minimum level of soil disturbance (reduced or zero tillage), (ii) permanent or semi-permanent organic soil cover, particularly through the retention of crop residues on the soil surface as mulch/cover, (iii) crop diversification in associations or sequences including a legume and (iv) integrated weed management. Recently, 'controlled traffic' that lessens soil compaction due to wheels of tractors is being considered as another component of CA.

Resource Conserving Technologies vs. Conservation Agriculture

In advancement of the concept of CA, the term 'resource conserving technologies' appears. CA-based RCTs provides the potential benefits on resource conservation, input-use efficiency, yield enhancement, soil health improvement and adaptation to changing climate. Recent estimates have revealed that extent of adoption of CA-based RCTs is around 3.9 M ha in South Asia (Jat *et al.*, 2011). Generally, RCTs and CA system are referred as the same but there is distinguish difference between both these concepts. Resource conserving technologies refer to those practices which enhance input/resource-use efficiency. In contrast, CA practices only refer to RCTs with minimum soil disturbance, permanent soil cover with residue and diversified crop rotation. For example, new varieties with improved nitrogen-use efficiency or zero/reduced tillage (without residue cover) systems that saves fuel and improve plot-level water-use efficiency may be considered as RCTs but not as CA. Further, it may be stated that not all RCTs are compatible with CA system, *e.g.* laser land levelling in presence of residue (Singh *et al.*, 2016). Some RCTs, while attractive in the near-term, may be unsustainable in the longer-term. For example, the use of zero tillage without residue retention and without suitable rotations under some circumstances, can be more harmful to agro-ecosystem productivity and resource quality than a continued conventional practices (Sayre, 2000).

Types of RCTs

Over the past few decades, rapid strides have been made to evolve and spread various approaches to enhance resource conservation. Various CA-based RCTs for crop management include zero and reduced tillage with residues recycling, laser assisted precision land levelling, direct drilling into the residues, direct-seeded rice, brown manuring with *Sesbania*, unpuddled mechanical transplantation of rice, raised bed planting system, crop diversification and associated component technologies like site-specific nutrient management (SSNM), leaf colour chart (LCC), green seeker, soil test crop response, fertigation, customized fertilizer and agro-forestry, which help in conservation of soil, water and nutrients. The RCTs approaches are different across environments. The approaches for Indo-Gangetic Plains (IGPs) are different from that of hill region or desert region. It also differs according to various types of cropping systems.

In hill farming systems, rain water harvesting on watershed basis for availing water for irrigation and other purposes, mulching to reduce run off and increasing water infiltration into the soil surface, contour farming and strip cropping *etc.*, are used under RCT approaches while for IGPs region furrow irrigated raised bed system (FIRBS), zero tillage, laser land leveling, use of different seeding machines *etc.*, are followed.

The major natural resources in agricultural production system are soil, water, vegetation and climate. Some of the modern tools and techniques/practices which have the potential for resource conservation and improving the use efficiency of these resources are discussed here.

Zero/Reduced Tillage

Zero tillage (ZT) was first started in India during 1980s. It was promoted by Imperial Chemical Industries for promoting paraquat herbicide for controlling weeds in zero-till wheat (Das *et al.*, 2016). Currently, this technology is being promoted in rice-wheat cropping systems (RWCS) of IGP because of its potential economic and environmental benefits. It is the concept of growing crops over the soil that is undisturbed from the tillage practices. In ZT, primary tillage is completely avoided and secondary tillage is done in the row zone only. Sowing or drilling is accomplished with a specially designed zero-till seed-drill/planter, which has inverted 'T'-type furrow opener to make a narrow slit in the soil for simultaneous placing of seed and fertilizer (Gupta and Seth, 2007).

Zero-tilled sowing of wheat after rice has been the most effective practice to date in the IGPs. Sowing of wheat with conventional method requires at least a week for preparatory cultivation (3-4 tillage operations) after harvesting of traditionally grown rice crop. Delay in sowing due to late harvesting of conventional rice results in substantial yield loss in wheat as the grain filling stage coincides with pre-monsoon hot weather condition. This practice (ZT) can advance sowing of wheat following traditional rice by 7-10 days and yield loss caused by delayed sowing can be minimized. With further refinement of ZT technology, the concept of 'double no-till' system has been developed where both rice and wheat crops are drilled with minimum soil disturbance (Gupta and Seth, 2007). This system of cultivation improves physical conditions of soil which facilitates desired tilth and timely sowing of the succeeding wheat crop.

Adoption of ZT in wheat provides the benefits in terms of saving in time, energy (fuel) and cost of cultivation. Further, retention of residues helps in reducing the pollution due to burning of straw. On an average, ZT results in reduction of cost of cultivation by 2,500-3,000/ha and yield advantages by 15-20 per cent (Das *et al.*, 2016). A study in Punjab in 2009-10 *rabi* season revealed that, adoption of ZT enabled the farmers to save 4,500/ha and 50 litre/ha of diesel, and thus reduced the emission of CO_2 to about 94.12 million tonnes (2.6 kg on burning of one litre of diesel) in the environment (Kaur and Singh, 2013). Another study in Haryana found that, ZT on an average saves around 60 litre fuel (diesel) per ha thus reducing emission of CO_2 by 156 kg/ha/year (Hobbs and Gupta, 2002; Gupta *et al.*, 2004). In ZT technology, retention of residues on soil surface lowers canopy temperatures by 1-1.5 °C during grain filling stage (cooling effect of transpiration) because of sustained soil moisture availability to the crops (Jat *et al.*, 2009).

The use of ZT is being considered as a new tool in integrated weed management. In this technology, the soil is left undisturbed. Therefore, weed seeds lying deeper in soil, do not get exposed to congenial environment for germination. Thus weed infestation in ZT is reduced over time. Continuous adoption of ZT for 3-4 years is reported to reduce the infestation of *Phalaris minor* (littleseed canary grass) in wheat. Thus, the load of herbicide use and risk of herbicide resistant *P. minor* can be avoided. Weeds are primarily controlled by using herbicides. For controlling weeds, glyphosate or paraquat @ 1.0 kg/ha may be applied 7-10 days before sowing (pre-plant treatment) (Das *et al.*, 2014).

Figure 17.2: Zero-Till Sowing by Happy Seeder in Rice Residue.

Machinery Developed

Several new types of zero-till seed-cum-fertilizer drill/planter [fitted with either a double disk openers or mechanical dibbler-'punch planter', shredders-spreader (Happy Seeder) or roto-coulter type disc-drills] have been developed to suit the varying conditions and requirements. Happy seeder/turbo seeder has been developed for direct drilling of crops in surface residue (loose and anchored) conditions. This drill/planter can operate in the surface residues load up to 10 t/ha. Similarly, there are other planters such as rotary-disc drill for direct drilling in residue conditions (Gupta and Seth, 2007; Pathak *et al.*, 2012). With the development and refinement of these implements, it has become easier to plant crops with through retained residue conditions.

The major benefits of ZT in wheat are summarized below (Hobbs and Gupta, 2002; Gupta and Seth, 2007; Gupta and Sayre, 2007; Erenstein and Laxmi, 2008; Jat *et al.*, 2009).

- Reduction in cost of cultivation through less use of fuels, herbicides, water for irrigation, labour *etc.*
- On an average, wheat sowing of can be advanced by at least one week, thereby reducing yield losses by 1-1.5 per cent per day after optimum wheat sowing time
- Timely planting of crop resulting in higher productivity
- Weed infestation is less and the crop does not lodge at maturity
- Savings of irrigation water up to 15-20 per cent, on an average
- 10-15 per cent more plant stand in ZT fields
- Improved input-use efficiency
- ZT helps in improving soil structure and sustaining its fertility, mainly by soil organic matter build-up through retention and decomposition of crop residues/stubbles
- Reduction in green-house gases (GHGs) emissions owing to less burning of crop residues. ZT provides direct mitigation benefits through carbon sequestration
- On an average, saves 50-60 litre/ha of fuel, thus reduced the emission of CO_2 in the environment
- Retention of crop residues on the soil surface reduces soil erosion and compaction. It also moderates the soil temperature and helps in proper crop growth. Soil aggregation (>0.25 mm) under ZT system was found to be higher than conventional planting system. These will lead to better root development and crop establishment
- ZT leads to less oxidation of organic carbon through reduced tillage and the activity of flora and fauna is increased.

Bed Planting

Bed planting facilitates the sowing of crops on raised beds (narrow or broad

Figure 17.3: Bed Planter.

beds) separated by furrows using a raised-bed planter. Raised beds are usually made at 60-100 cm apart where 2-3 rows of crops are sown and irrigation is given in the furrows. First time raised beds are prepared after conventional tillage using a raised-bed planter and are left for 30 days to settle. These beds can be used for direct drilling for the subsequent years, but with little reshaping of the beds once in a year, preferably before *kharif* crops are planted. In RWCS of western IGPs, raised beds are mare at 37 cm wide at the top and 15 cm in height and separated by furrows having a 30-cm-wide top. The distance between the centers of two subsequent furrows is kept at 67 cm. Rice is drilled on the raised beds keeping two rows on both sides at the top of the beds with a spacing between two rows at 25 cm. After rice, wheat is directly sown using the same bed planter that reshapes the beds along with simultaneous placement of seeds and fertilizer (Bhushan *et al.*, 2007; Jat *et al.*, 2009; Das *et al.*, 2014).

Bed planting facilitates more efficient use of water under rainfed as well as irrigated situations owing to efficient storage of water and drainage of excess water. It helps in conserving inputs like seed, water, nutrient, *etc.* Bed planting has the potential to save water by up to 30-50 per cent in wheat (Hobbs and Gupta, 2002). The furrow irrigated raised-bed system (FIRBS) of wheat cultivation results in savings of inputs like seed by 25-40 per cent, water by 25-40 per cent and nutrients by 25 per cent, without compromising with the crop productivity. It also provides the option for crop diversification (Jat *et al.*, 2005; Pathak *et al.*, 2012). The other benefits include: (i) less lodging, (ii) less soil compaction and improved soil aeration, (iii) safe disposal of excess water specially in heavy textured soils, (iv) more potential tillers and better canopy formation, (v) more interception of solar energy by crop canopy, (vi) reduced weed growth on the beds, (vii) higher efficiency of chemical

Figure 17.4: Bed Planted Wheat.

weed control and easier mechanical weeding on the furrows, (vii) moderation of soil temperature, and (viii) better crop growth and productivity (up to ~20 per cent).

Some precautions must be taken for adopting bed planting technique. These include: (i) furrows should be filled with irrigation water up to 3/4th height of the raised beds and water should not flow over top of the beds, (ii) the field should be clod-free and uniformly levelled, preferably be laser levelled, (iii) beds should be made weed-free before sowing of each crop using herbicides like glyphosate or paraquat under non-cropped situation, (iv) proper soil moisture should be maintained in the beds for better germination, (v) water can be applied at the appearance of hairline cracks on the soil surface at the bottom of the furrows.

Laser Land Levelling

In India, surface irrigation (mainly flood irrigation) is a common practice wherein a significant amount of irrigation water is lost during application because of poor management and uneven fields that lead to lower crop yields, higher irrigation costs and poor input-use efficiency (Jat *et al.*, 2009). Laser assisted precision land levelling (commonly known as laser land levelling) is a pre-requisite technology for adopting RCTs like zero tillage or bed planting that require properly levelled fields for uniform distribution of inputs like seeds, water and fertilizer. It uniformly levels the land surface (± 2 cm) and alters the fields to create a constant slope of 0 to 0.2 per cent using laser equipped drag scraper. This practice makes use of large horsepower tractors and soil movers that are equipped with global positioning systems (GPS) and/or laser-guided instrumentation so that the soil can be moved either by cutting or filling to create the desired slope/level (Pathak *et al.*, 2012).

Figure 17.5: Laser Land Leveller.

Precision land levelling provides a very accurate, smooth and levelled field which facilitates application of less water more uniformly under surface irrigated systems and reduces leaching losses, thus improving the use-efficiency of applied water and reducing irrigation cost. This can also reduce water losses by suppressing evaporative and percolation losses owing to faster irrigation times (as irrigation water reaches to the tail ends of the fields in less time) and elimination of depressions.

This technology helps in saving of irrigation water by up to 25-30 per cent and improves the use-efficiency of applied N. There is about 3-4 per cent increase in the area irrigated due to removal and/or fewer bunds and channels. Other benefits include improved crop stand and productivity (up to 20 per cent) and reduced labour requirement (up to 35 per cent). This technique can be very useful for rice since irrigation water requirement is reduced drastically (Gill *et al.*, 2002; Jat *et al.*, 2009; Das *et al.*, 2014).

Direct-Seeded Rice

Direct-seeded rice (DSR) is an emerging production system as consequences of water scarcity, labour scarcity, energy crisis and continuous climate change. In DSR system, rice can be sown directly on dry seed bed (dry-DSR) or pre-germinated seeds can be sown on puddled soil (wet-DSR) or sowing can be done into shallow standing water (water seeding). This can save inputs like water, labour, fuel and time. Dry-DSR is grown under aerobic soil conditions without prolonged periods of flooding as periodic irrigations maintain the soil at field capacity. It avoids huge water and labour requirement during land preparation (puddling) and/or transplanting, thereby reducing overall water and labour demand. Direct-seeded rice significantly lowers total water requirement (up to 35-57 per cent) and labour requirement (up to 67 per cent) than transplanted rice depending on season, location, and type of DSR (Farooq *et al.*, 2011; Mahajan *et al.*, 2013). Moreover, rice cultivation through puddling and transplanting deteriorates the soil structure, forms a hard pan which may suppress the growth and yield of succeeding crop. DSR maintains good soil physical conditions which facilitates desired tilth and timely sowing of the succeeding crop.

DSR can fetch yield similar to transplanted rice if weeds are managed properly. Weed infestation is considered as the main biological constraint to the success of DSR. High weed pressure and shift toward more difficult-to-control weed flora limit the productivity. It even causes complete crop failure under extreme cases (Rao *et al.*, 2007). Judicious use of herbicides (either in a sequence or mixture) along with other cultural practices provides an effective weed control and good yield.

Direct seeding has a number of other benefits over transplanting of rice such as faster and easier planting, reduced drudgery, improved physical conditions of soil, efficient use of irrigation water, higher tolerance to water deficit, lesser emissions of methane, and often higher profit in areas with assured water supply (Kumar and Ladha, 2011; Mahajan *et al.*, 2013). In addition to this, DSR matures 7-10 days earlier than the transplanted rice which allows timely planting of the succeeding crop *rabi* season crop especially wheat in IGPs.

Brown Manuring

Brown manuring (BM) is a practice of concurrent sowing of Dhaincha (*Sesbania*) in alternate rows of rice manually and subsequent application of 2, 4-D @ 0.25-0.5 kg/ha to knockdown the *Sesbania* at 25-30 days after sowing. Rice is drilled in lines and *Sesbania* is broadcasted in the inter-row spaces on moist soil. Bispyribac-Na, a broad-spectrum herbicide can be applied (20-25 g/ha) instead of 2, 4-D. Brown manuring has the potential to offset the drawbacks of green manuring which requires additional water for irrigation, fuel costs for its incorporation. And also little water availability during peak summer season hinders to take the full benefits of green manuring (Pathak *et al.*, 2012).

Sesbania while growing with rice provides a quick vegetative cover and smothers various weed species and conservers soil moisture by checking evaporation from the soil surface. After knocking down, it forms a brown mulch on the soil surface and thus suppresses weed, reduces herbicide use and supplies about 15-20 kg N/ha with a fresh biomass of 10-12 t/ha (Das *et al.*, 2016). It conserves soil moisture, lowers irrigation requirement, improves soil health and enhances grain yield and profitability without adding much on cost of production (only seed cost).

This technique can also be adopted for other cereal crops like maize, pearl millet, sorghum, *etc.* where *Sesbania* can be cut manually and spread in between the crop rows as mulch for conservation of resources (Das *et al.*, 2014).

Crop Residue Management

Crop residues are the parts of a plant or crop which are left in the field after harvest or discarded during processing. Management of crop residues or efficient recycling of crop residues is crucial, especially in rice-wheat belt of IGPs, where it is burnt *in situ*. India produces huge amount of crop residues (more than 600 million tonnes) annually of which major quantity is contributed by rice and wheat. A vast potential is available to effectively recycle the huge volume of crop residues.

A large quantities of residues are burnt *in situ* every year by the farmers to facilitate land preparation for crop succeeding crop. Crop residue on burning increases rates of mineralization that rapidly depletes nutrients and organic matter from the soil and also causes air pollution. It is estimated that burning of one tonne of straw releases 3 kg particulate matter, 60 kg CO, 1460 kg CO_2, 199 kg ash and 2 kg SO_2 (Gupta *et al.*, 2004). Various new drills/planters such as Happy/turbo seeder or rotary-disc drill have been developed which are able to cut through the crop residue for zero-tillage crop planting and can operate in residue load up to as much as 10 t/ha. Thus, burning of residue can be avoided and release of some 13-14 tons of carbon dioxide potentially can be reduced (Gupta *et al.*, 2004; Gupta and Seth, 2007).

In CA, a critical level of crop residue as permanent or semi-permanent organic soil cover must not kept on the soil surface. Retention of residue on the surface provides numerous benefits which are listed below.

- It protects organic matter enriched top soil from weather aggressions and water erosion by reducing crust formation and enhanced infiltration
- Organic matter formed after residue decomposition acts as a reservoir of plant nutrients, prevents nutrient leaching and improves cation exchange capacity (CEC)
- Residue left on the soil surface helps in suppression of germination and growth of weeds, thereby reducing herbicide use
- Crop residue maintains or improves soil physical, chemical and biological properties and prevents land degradation. It helps in water conservation through enhanced water infiltration and reduced evaporation
- Retention of crop residue moderates soil temperature by raising minimum soil temperature in the winter and decreasing soil temperature during summer
- It provides climate change mitigation benefits through carbon sequestration
- It provides shelter and food for the soil biota and enhances microbial and enzymatic activity
- It plays an important role ameliorating soil acidity and alkalinity
- Crop residue is also an important source of fodder for animals

However, rate of decomposition and release of nutrients from residue depend on soil, climatic conditions and the C: N ratio residue. Nevertheless, there are some negative aspects of crop residue retention, like increase in incidence disease and pest infestation with retained residue which may hinder its adoption. But the benefits appear to be greater than the disease-pest factor.

Crop Diversification

Crop diversification refers to a shift from existing sole cropping or monoculture towards more balanced and diversified cropping system. Different planting and harvesting times can reduce risks from weather abnormalities and disease-pest infestations and new crops can be adjusted in the existing cropping systems. Therefore, crop diversification has been recognized as an effective strategy for (i) reducing risks associated with yield, market and prices (ii) reducing the degradation of natural resources and the environment (iii) attaining national goals like providing food security and nutritional security, employment generation, income growth, poverty alleviation and earning foreign exchange (iv) judicious use of land and water resources, improving biodiversity, soil health, agro-ecosystem and environment (Hegde *et al.*, 2003; Reddy and Suresh, 2009).

Short duration crops like pulses, oilseeds and other high value crops can be used as sequential or intercrops for crop intensifications and higher yield stability. Thus, an increase in cropping intensity can contribute substantially to additional demands of food and cash crops. Development of new crop varieties with enhanced photosynthetic efficiency and shorter duration would be of massive help in increasing cropping intensity. Pigeon pea, one of the most important pulses in south Asia has shown tremendous potential for diversification of rice crop in the IGPs. Introduction of extra short duration pigeon pea (ICPL-88039) and FIRBS planting technique in this region would enable small and marginal farmers to utilize limited land and water resources in more efficient manner (Ramesh *et al.*, 2016).

There are two approaches of crop diversification. These are:

- **Horizontal diversification:** It takes place through substitution or intensification (addition) of the cropping system for improving the system productivity.
- **Vertical diversification:** In this approach, the economic produce of various crops is refined and graded to add additional values to the final product.

Site-Specific Nutrient Management

Nutrient management and recommendation process synchronized with crop demand or plant need-based application is crucial for minimizing losses from the soil-plant system and achieving high yield and nutrient-use efficiency. Site-specific nutrient management (SSNM) is an approach for providing need based feeding of crops with nutrients while recognizing the inherent spatial variability. The SSNM emphasizes on principles of '5Rs', *i.e.* the right source, the right rate, the right time, the right place and the right method. It takes into account all pathways of plant nutrient flows, and judiciously makes use of all combinations of fertilizers, bio-fertilizers, organic manures, crop residues and nutrient efficient genotypes for sustaining agricultural production. It avoids indiscriminate use of fertilizers and aims at nutrient supply at optimum rates and times to attain improved productivity, profitability and nutrient-use efficiency by the crop (Ramesh *et al.*, 2016). The SSNM approach comprises three steps, *viz.* (i) establishing attainable yield targets (ii) effectively use existing nutrient sources and (iii) application of fertilizers to fill the deficit between demand and supply of nutrients.

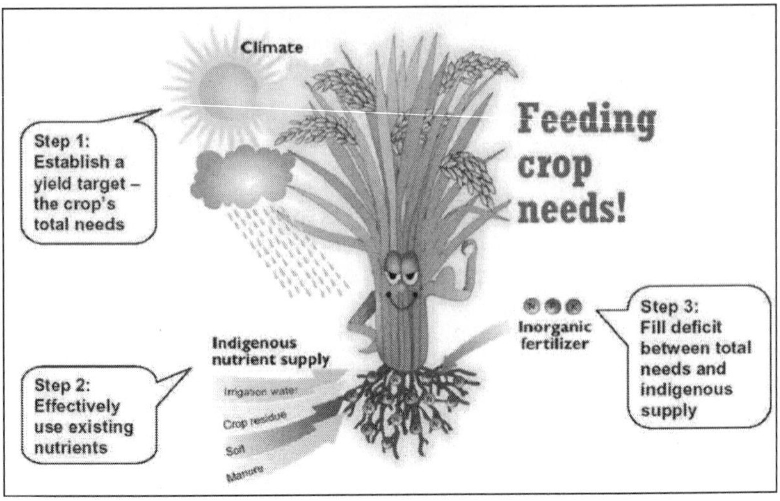

Figure 17.6: Site-Specific Nutrient Management (*Source*: www.irri.org/irrc/ssnm).

Chlorophyll Meter

The chlorophyll meter, also known as SPAD (Soil plant analysis development) meter, is a simple, non-destructive and portable diagnostic tool for instant estimation of leaf nitrogen (N) status. Nitrogen content is related to chlorophyll content, which is displayed in arbitrary values (0-99.9) in SPAD meter. It has been used successfully for synchronizing N application with N needs of crops (plant need-based N management) and improving N-use efficiency. Wheat crop showed response to N application at maximum tillering stage at SPAD value of below 44. Application of 30 kg N/ha at critical SPAD value of 42 at maximum tillering stage increased the wheat grain yield by up to 20 per cent. It has been found that plant need-based N management through chlorophyll meter reduces N requirement of rice without any decrease in yield (Singh *et al.*, 2002).

Leaf Colour Chart

The cost of the chlorophyll meter restricts its widespread use by farmers. Leaf colour chart (LCC) is a simple hand-held and inexpensive tool for quick estimation of plant N status without causing damage. Use of LCC promotes timely, efficient and plant need-based N management, especially in rice and wheat (Pathak *et al.*, 2012). Leaf colour chart is a high quality plastic strip with different shades of green colour ranging from light yellowish green to dark green (Figure 17.7). Japanese scientists first developed this N management tool (Furuya, 1987). A LCC uses the green color intensity of rice leaves to assess the N status, since leaf color intensity is directly related to leaf chlorophyll content and leaf N status. (IRRI-CREMNET, 1998). It is a promising eco-friendly tool and now is being standardized with chlorophyll meter.

The LCC has been standardized and refined over the years. The new LCC has 4 colour strips by removing panels 1 and 6 (pale yellow and dark green) of the 'IRRI-PhilRice' version of LCC (as these colours were rarely seen under field conditions).

Figure 17.7: Leaf Colour Chart (LCC).

It has reduced cost but the effectiveness remains the same. It is light-weight, smaller (can easily fit in pocket) and has a provision for a neck sling (Figure 17.8).

Leaf colour chart can help farmers avoid over application of N to rice. Nitrogen applied as per LCC values can save 20-30 kg of fertilizer N/ha without compromising with rice yield. For achieving best results of LCC-based N management, N should be applied (20-30 kg/ha) at LCC<3 in basmati and direct-seeded rice and at LCC<4 in coarse and hybrid rice (Pathak *et al.*, 2012). Moreover, in LCC-based N management, basal application of N can be skipped without any reduction in terms of grain yield and fertilizer N-use efficiency.

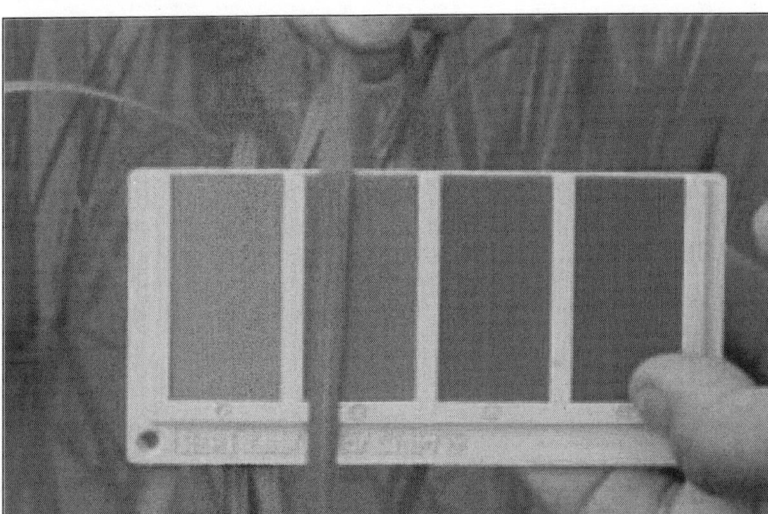

Figure 17.8: Leaf Colour Chart (LCC).

Some precautions must be taken into account for LCC-based N management. These are: (i) the topmost, fully expanded and healthy leaf should be taken to compare with the LCC, (ii) middle part of the leaf must be placed on LCC colour strips for comparison, (iii) comparison must be done during the morning hours and (iv) LCC must not be exposed to direct sunlight during the comparison

Future Thrusts

The growing concern of productivity, resource conservation and environmental degradation is now the driving forces for adoption of RCTs. CA based RCTs offer a new paradigm for agricultural research and development which mainly aim at sustainable agricultural development. However, for realizing the full potential benefits of these technologies some key perspectives and future researchable issues have been identified. These are as follows:

- ☆ Evaluation of long-term impacts of various CA based RCTs on crop, soil health, biodiversity, resource-use efficiency, C sequestration, GHGs emission, climate and ecosystem services in various production systems and agro-ecologies for enhanced profitability and sustainability
- ☆ Development of new implements for ZT and other RCTs requiring low power especially under small farm conditions
- ☆ Developing and promoting promising CA practices for each agro-ecological region, particularly in rainfed and dryland agro-ecosystems
- ☆ Identification and standardization of region-specific new cropping, intercropping and novel farming systems for crop diversification through substitution or intensification
- ☆ Discouraging burning of crop residues and developing new low-cost technologies for *in situ* management of crop residues for improving soil health and reducing environmental pollution.
- ☆ Identification and developing new crop varieties (such as shorter duration) with high resource-use efficiency for efficient management of soil and water
- ☆ Identification and developing various SSNM technologies for enhancing nutrient-use efficiency and reducing nutrient losses and pollution hazards
- ☆ Development of economically and ecologically viable strategies for weed, nutrient, pest and disease management
- ☆ Increasing the number of on-farm research and demonstration trials and improvement in transfer of technology that can accelerate technology adoption

Conclusion

Keeping in view the deteriorating natural resource bases, RCTs such as zero and reduced-tillage systems, better crop residue management and planting systems, have been evolved over the past few decades for the efficient management of soil, water and nutrient. Resource conserving technologies have the potential to improve the sustainability of agricultural production system and improve natural resource bases. These, further improve the quality of environment by mitigating GHGs emissions and help in adapting to climate change. However, these technologies are highly site-specific. Hence, various RCTs across different production systems, agro-ecologies, cropping seasons and farm conditions need to be developed and their cost-effectiveness must be validated to have significant adoption, especially under Indian conditions, where mainly marginal and small farm holdings (up to 2 ha) predominate. It is, therefore, need of the time to adopt and further disseminate the resource conserving technologies that improve the sustainability of agricultural production system without compromising with the productivity.

REFERENCES

Abrol, I.P. and Sangar, S. 2006. Sustaining Indian agriculture-conservation agriculture the way forward. *Current Science* **91**(8): 1020-2015.

Anonymous, 2015. Vision 2050. Indian Agricultural Research Institute, New Delhi–110012, pp. 1-37.

Bhushan, L., Ladha, J.K., Gupta, R.K., Singh, S., Padre, A.T., Saharawat, Y.S., Gathala, M. and Pathak, H. 2007. Saving of water and labour in a rice-wheat system with no-tillage and direct seeding technologies. *Agronomy Journal* **99**: 1288-1296.

Das, T.K., Sepat, S., Singh, P., Bhattacharyya, R. and Pathak, H. 2014. Conservation agriculture for enhancing crop productivity and resource-use efficiency. *Indian Farming* **64**(1): 18-21.

Das, T.K., Sharma, A.R., Rana, D.S. and Paul, T. 2016. Conservation agriculture. In: *Modern Concepts of Agronomy* (Rana, D.S., Ghosh, P.K., Shivay, Y.S. and Singh, G. Eds.). pp. 62-86. Indian Society of Agronomy, New Delhi.

Derpsch, R. 2001. Frontiers in conservation tillage and advances in conservation practice. In: *D.E. Stott, R.H. Mohtar and G.C. Steinhardt (Eds). Sustaining the Global Farm. Selected papers from the 10th International Soil Conservation Organization Meeting*. May 24-29, 1999 at Purdue University and the USDA-ARS National Soil Erosion Research Laboratory, pp. 248-254.

Erenstein, O. and Laxmi, V. 2008. Zero tillage impacts in India's rice-wheat systems: A review. *Soil and Tillage Research* **100**: 1-14.

FAO, 2017. Food and Agriculture Organization of the United Nations. Corporate Statistical Database. Available online at http: //www.fao.org/ag/ca/6c.html. Accessed on December, 2017.

Farooq, M. and Siddique, K.H.M. 2015. Conservation agriculture: Concepts, brief history, and impacts on agricultural systems. In: *Conservation Agriculture*. (Farooq, M. and Siddique, K.H.M. Eds.). pp. 3-17. Springer International Publishing, Switzerland.

Farooq, M., Siddique, K.H.M., Rehman, H., Aziz, T., Lee, D.J. and Wahid, A. 2011. Rice direct seeding: experiences, challenges and opportunities. *Soil and Tillage Research* **111**(2): 87-98.

Furuya, S. 1987. Growth diagnosis of rice plants by means of leaf color. *Japan Agricultural Research Quarterly* **20**: 147-153.

Gill, M.A., Chaudhary, M.A., Ahmed, M. and Mujeeb-ur-Rehman, A. 2002. Water management, cultural practices and mechanization. In: *National Workshop on Rice-wheat Cropping System Management* (Akhtar, M.S. and Nabi, G. Eds.). Pakistan Agricultural Research Council, Islamabad, Pakistan, 11-12 December, 10p.

Gupta, P.K., Shivraj, S., Nahar, S., Dixit, C.K., Singh, D.P., Sharma, C., Tiwari, M.K., Gupta, R.K. and Garg, S.C. 2004. Residue burning in rice-wheat cropping system: causes and implications. *Current Science* **87**(12): 1713-1717.

Gupta, R. and Seth, A. 2007. A review of resource conserving technologies for sustainable management of the rice-wheat cropping systems of the Indo-Gangetic plains (IGP). *Crop Protection* **26**: 436-447.

Gupta, R., Sayre, K. 2007. Conservation agriculture in South Asia. *Journal of Agricultural Sciences Cambridge* **145**(3): 207-214.

Hegde, D.M., Prakash Tiwari, S. and Rai, M. 2003. Crop diversification in Indian Agriculture. *Agricultural Situation in India*. August, 2003. pp. 255-272.

Hobbs, P.R. and Gupta, R. 2002. Resource conserving technologies for wheat in the rice-wheat system. In: *Improving the Productivity and sustainability of rice-wheat systems: Issues and impacts* (Ladha, X. Ed.). ASA Special Publication. ASA, Madison, WI, USA.

IRRI-CREMNET, 1998. IRRI-Crop and Resource Management Network. Progress report for 1997. IRRI, Manila, Philippines.

Jat, M.L., Gathala, M.K., Ladha, J.K., Saharawat, Y.S., Jat, A.S., Kumar, V., Sharma, S.K., Kumar, V. and Gupta, R. 2009. Evaluation of precision land leveling and double zero-till systems in the rice–wheat rotation: Water use, productivity, profitability and soil physical properties. *Soil and Tillage Research* **105**: 112-121.

Jat, M.L., Gupta, R., Saharawat, Y.S. and Khosla, R. 2011. Layering precision land leveling and furrow irrigated raised bed planting: Productivity and input use efficiency of irrigated bread wheat in Indo-Gangetic Plains. *American Journal of Plant Sciences* **2**(4): 578-588.

Jat, M.L., Singh, S., Rai, H.K., Chhokar, R.S., Sharma, S.K. and Gupta, R.K. 2005. Furrow irrigated raised bed planting technique for diversification of rice-wheat system of Indo-Gangetic Plains. *Journal of Japan Association for International Cooperation for Agriculture and Forestry* **28**(1): 25-42.

Kaur, R. and Singh, A. 2013. Resource conservation technologies for sustainable agriculture. *Indian Farming* **62**(12): 13-20.

Kumar, V. and Ladha, J.K. 2011. Direct Seeding of Rice: Recent Developments and Future Research Needs. *Advances in Agronomy* **111**: 297-413.

Mahajan, G., Chauhan, B.S. and Gill, M.S. 2013. Dry-seeded rice culture in Punjab State of India: Lessons learned from farmers. *Field Crops Research* **144**: 89-99.

Pathak, H., Sharma, A.R., Das, T.K. and Jat, M.L. 2012. Adaptation and mitigation of climate change with conservation agriculture. In: *Climate Change Impact, Adaptation and Mitigation in Agriculture: Methodology for Assessment and Application* (Pathak, H., Aggarwal, P.K. and Singh, S.D. Eds.). pp. 222-240. IARI, New Delhi.

Ramesh, Negi, S.C. and Rana, S.S. 2016. Resource conservation technologies (RCTs) – needs and future prospects: a review. *Agricultural Reviews* **37**(4): 257-267.

Rao, A.N., Johnson, D.E., Sivaprasad, B., Ladha, J.K. and Mortimer, A.M. 2007. Weed management in directseeded rice. *Advances in Agronomy* **93**: 153-255.

Reddy, B.N. and Suresh, G. 2009. Crop diversification with oilseed crops for-maximizing productivity, profitability and resource conservation. *Indian Journal of Agronomy* **54**(2): 206-214.

Sayre, K. 2000. Effects of tillage, crop residue retention and nitrogen management on the performance of bed-planted, furrow irrigated spring wheat in northwest Mexico. Paper presented at the Conference of the International Soil Tillage Research Organization, 15; Fort Worth, Texas, USA; 2-7 Jul, 2000.

Singh, B., Singh, Y., Ladha, J.K., Bronson, K.F., Balasubramanian, V., Singh, J. and Khind, C.S. 2002. Chlorophyll meter–and leaf color chart–based nitrogen management for rice and wheat in Northwestern India. *Agronomy Journal* **94**(4): 821-829.

Singh, V.K., Mishra, R.P., Ravishankar, N. and Rani, M. 2016. Resource use efficiency. In: *Modern Concepts of Agronomy* (Rana, D.S., Ghosh, P.K., Shivay, Y.S. and Singh, G. Eds.). pp. 37-61. Indian Society of Agronomy, New Delhi.

Conservation Agriculture Technologies *Pages* **211–223**
Editor: **Rahul Kumar, Abhinandan Singh and Pankaj Kumar Ojha**
Published by: **BIOTECH BOOKS, NEW DELHI**

18 Conservation Agriculture: An Overview

Vipin Kumar[1] *and Ajay Kumar*[2]

[1]*Department of Plant Pathology, S.V.B.P.U.A. and T., Meerut – 250 110, Uttar Pradesh*
[2]*Department of Plant Pathology, N.D.U.A. and T., Faizabad – 224 229, Uttar Pradesh*

Background

In 2000, about 850 million people were suffering from hunger, 815 million of which in developing countries. From a global perspective, hunger is a problem not of food production but of accessibility (FAO, 2006). Although recent predictions point to reduced population growth, a considerable increase in overall production is still required a major challenge is land and water resources. Future yield increases limited compared to past trends but by 2030, food production must double to keep pace with demand. Sub-Saharan Africa remains a hunger hot spot with a stagnating trend in per capita food production, while other regions are increasing (FAO, 2005).

Countries with a large population like China and India, and which in the past achieved food self sufficiency, now face the challenge of maintaining and increasing high yield levels. Agricultural land available per capita is expected to decline (FAO, 2002) while revolutionary technologies for significantly higher production potential. Water is one of the most precious natural resources for agricultural production and accounts for 70 per cent of water use (FAO, 2002). It is predicted that by 2025 water consumption will exceed "blue water" availability if current trends continue (Ragab and Prudhomme, 2002).

In Indian state of Punjab, characterized by intensive irrigated agriculture, the groundwater table is falling at a rate of 0.7 m per year (Aulakh, 2005). Soil affects not only production, but also the management of other natural resources, such as water. Soil structure is strongly correlated with the organic matter content and the soil life. Organic matter stabilizes soil aggregates, provides feed to soil life and acts as a sponge for soil water. The lack of yield response to high doses of fertilizer in the Indo-Gangetic Plains can be attributed to poor soil health resulting from overexploitation (Aulakh, 2005). In the Indian states of Uttarakhand and Haryana, the organic carbon content in soils minimum values less than 0.1 per cent (PDCSR, 2005).

Introduction

Attaining food security for a growing population while sustaining agricultural systems under the current scenario of depleting natural resources, negative impacts of climatic variability, spiralling cost of inputs and volatile food prices are the major challenges before most of the Asian countries. In addition to these challenges, the principal indicators of non-sustainability of agricultural systems includes: soil erosion, soil organic matter decline and salinization.

Mainly caused by:

(i) Intensive tillage induced soil organic matter decline, soil structural degradation, water and wind erosion, reduced water infiltration rates, surface sealing and crusting and soil compaction.

(ii) Insufficient return of organic material.

(iii) Monocroping.

Conservation agriculture (CA) is a resource-saving agricultural production intensification and high yields while enhancing the natural resource with three interrelated principles, along with other good production practices of plant nutrition and pest management. Traditional agriculture based on tillage and highly mechanized responsible for soil erosion problems, surface and undergroundwater pollution and more water consumption (Wolff and Stein, 1998). Hence, conservation agriculture (CA) is cultivate annual and perennial crops based on no vertical perturbation of soil (zero and conservation tillage), with crop residue management and cover crops, in order to offer a permanent soil cover and a natural increase of organic matter content in surface horizons. Soil erosion is highly reduced soil agronomic inputs transport slightly reduced, while pesticide bio-degradation is enhanced. It protects surface and groundwater resources from pollution and also mitigates negative climate effects. Hence, CA provides excellent soil fertility and also saves money, time and fossil-fuel. It is an efficient alternative to traditional agriculture, attenuating its drawbacks.

Definition

Conservation Agriculture is a system of integrated management of soil, water and biological resources combined with external inputs. As a concept of sustainable agricultural production, it is promoted by several organizations around the world, including FAO.

Objective of Agriculture

Its objective is to improve agriculture production by adopting economically, ecologically and socially sustainable methods. Conservation Agriculture aims to conserve, improve and make more efficient use of natural resources. It contributes environmental conservation as well as to enhanced and sustained agricultural production for farmers.

What is Conservation Agriculture?

It is a way of farming that conserves, improves, and ensures efficient use of natural resources. It aims to help farmers achieve profits with sustained production levels while conserving the environment. Thus, CA is a concept for optimizing crop yields, and economic and environmental benefits.

Key of CA

The key elements of CA include no-tillage, adequate retention of crop residues on the soil surface for mulching, innovative cropping systems and measure to reduce soil compaction through controlled traffic. Freewheeling of the farm machinery in moist fields creates ruts and compacts soil. This must be avoided if zero-till is to be practised for a longer time. CA practices have been widely adopted in tropics/ subtropical and temperate regions of the world for rain-fed and irrigated systems. Conservation agriculture is increasing steadily worldwide to cover about 108 m ha globally (7 per cent of the world arable land area). Please note the distinction between Conservation Tillage (CT) and Conservation Agriculture. CT only refers to tillage practices that leave at least 30 per cent of the soil surface covered by residues. Soil Conservation (SC), Conservation Tillage (CT) and Resource Conservation Technologies (RCTs) are not synonymous with Conservation Agriculture.

Principles of Conservation Agriculture

Conservation agriculture practices perused in many parts of the world are built on ecological principles making land use more sustainable (Wassmann, 2009; Behera *et al.*, 2010; Lal, 2013). Adoption of CA for enhancing Resource use efficiency (RUE) and crop productivity is the need of the hour as a powerful tool for management of natural resources and to achieve sustainability in agriculture. Conservation agriculture basically is three principles, which are linked and must be considered together for appropriate design, planning and implementation processes.

(i) Minimal Soil Disturbance

The soil biological activity produces very stable soil aggregates as well as various sizes of pores, allowing air and water infiltration. This process can be called "biological tillage" and it is not compatible with mechanical tillage. With mechanical soil disturbance, the biological soil structuring processes will disappear. soil disturbance maintains by optimum proportions of respiration gases in the rooting-zone, moderate organic matter oxidation, porosity for water movement, retention and release and limits the re-exposure of weed seeds and their germination (Kassam and Friedrich, 2009).

(ii) Permanent Organic Soil Cover

Soil cover is important to protect the soil against the deleterious effects of exposure to rain and sun provide micro and macro organisms in the soil with a constant supply of "food" alter microclimate in the soil for optimal growth and development of soil organisms including plant roots. In turn it improves soil aggregation, soil biological activity and soil biodiversity and carbon sequestration (Ghosh *et al.*, 2010).

(iii) Diversified Crop Rotations

The rotation of crops is not only necessary to offer a diverse "diet" to the soil micro organisms, but also for exploring different soil layers for nutrients that leached to deeper layers that can be "recycled" by crops in rotation. Furthermore, a diversity of crops in rotation leads to soil flora and fauna. Cropping sequence and rotations involving legumes helps in minimal rates of build-up population of pest species life cycle disruption biological nitrogen fixation, control of off-site pollution and enhancing biodiversity (Kassam and Friedrich, 2009).

Agriculture Conservation Status

Globally, CA is being practiced on about 125 M ha. The major CA practicing countries are USA (26.5 M ha), Brazil (25.5 M ha), Canada (13.5 M ha) and Australia (17.0 M ha). In India, CA adoption is still in the initial phases. Over the past few years, adoption of zero tillage and CA has expanded to cover about 1.5 million hectares (Jat *et al.*, 2012). The major CA based technologies adopted is zero-till (ZT) wheat in the rice-wheat (RW) system of the Indo-Gangetic plains (IGP). In other crops and cropping systems, the conventional agriculture based crop management systems are gradually undergoing a paradigm shift from intensive tillage to reduced/zero-tillage operations.

In India, efforts to adopt and promote conservation agriculture technologies have been underway for nearly a decade but it is only in the last 8 - 10 years that the technologies are rapid acceptance by farmers. Efforts to develop and spread conservation agriculture have been made through the combined efforts with several State Agricultural Universities, ICAR institutes. The spread of technologies is taking place in India irrigated regions in the Indo-Gangetic plains where rice-wheat cropping systems. The focus of developing and promoting conservation technologies has been on zero-till seed-cum fertilizer drill for sowing of rice-wheat system. Other interventions like as raised-bed planting systems, laser equipment aided land levelling, residue management practices, alternatives to rice-wheat system *etc.* It has been reported that the area planted with wheat adopting the zero-till drill has been increasing rapidly (Sangar *et al.*, 2005), and presently 25 per cent - 30 per cent of wheat is zero-tilled in rice-wheat growing areas of the Indo-Gangetic plains of India. In addition, raised-bed planting and laser land leveling are also being increasingly adopted by the farmers of the north-western region.

Its Advantages

Building up the soil organic matter also retains nutrients and improves the micro-flora in the soil. Soil organic matter can hold many times its weight in water building up the organic matter in the soil results in greater water retention. This reduces labour and fuel use and also reduces irrigation water. Through crop rotation, organic matter is placed in different soil strata, thereby making the soil more fertile.

Its Disadvantages

In general, farmers need to be more careful about timing of agricultural operations under Conservation Agriculture. Special attention has to be paid to weed control through hand weeding or by judicious use herbicides. However, farm production tends more stable than under conventional agricultural methods. So far there have been no pest problems that cannot to be overcome in Conservation Agriculture.

Table 18.1: Distinguishing Features of Conventional and Conservation Agriculture

Conventional Agriculture	*Conservation Agriculture*
Cultivating land, using science and technology to dominate nature	Least interference with natural processes
Excessive mechanical tillage and soil erosion	No-till or drastically reduced tillage (biological tillage)
High wind and soil erosion	Low wind and soil erosion
Residue burning or removal (bare surface)	Surface retention of residues (permanently covered)
Water infiltration is low)	Infiltration rate of water is high
Use of *ex-situ* FYM/composts	Use of *in-situ* organics/composts
Green manuring (incorporated)	Brown manuring/cover crops (surface retention)
Kills established weeds but also stimulates more weed seeds to Germinate	Weeds are a problem in the early stages of adoption but decrease with time
Mono cropping/culture, less efficient rotations	Diversified and more efficient rotations
Poor adaptation to stresses, yield losses greater under stress conditions	More resilience to stresses, yield losses are less under stress conditions
Productivity gains in long-run are in declining order	Productivity gains in long-run are in incremental order

Source: Sharma *et al.*, 2012.

Prospects of Conservation Agriculture

The direction that Asian countries take to meet their food and energy needs during the coming decades will have profound impacts on natural resource bases, global climate change and energy security for India, Asia and the world. These challenges draw attention to the need and urgency to address options by which threats to Indian/Asian agriculture due to natural resource degradation, escalating production costs and climate change can be met successfully. Asian farmers/ researchers will continue need assistance to reorient their agriculture and practices for producing more with less cost. Conventional agriculture practices do not seem to sustainable option for suitable gains in food-grain production and hence CA-based crop management solutions.

Indian/Asian context has the following prospects:

(i) **Reduction in crop cost production**: Many factor contributing to rapid adoption of zero-till technology. Most studies showed that the cost of wheat production is reduced by Rs. 2,000 to 3,000 ($ 33 to 50) per hectare (Malik *et al.*, 2005; RWC-CIMMYT, 2005). Cost reduction is attributed to savings on account of diesel, labour and input costs and particularly herbicides.

(ii) **Reduced incidence of weeds**: Most indicate reduced incidence of *Phalaris minor*, a major weed in wheat, when zero-tillage is adopted resulting reduced use of herbicides.

(iii) **Increased yields**: In managed zero-till planted wheat yields compared to traditionally prepared fields.

CA enhance the yield level of crops due to associated effects like prevention of soil degradation improved soil fertility improved soil moisture (increased rain water infiltration water holding capacity and reduced evaporation loss) and crop rotational benefits. Yield increases as high as 200–500 kg ha^{-1} with no-till wheat compared to conventional wheat under a rice-wheat system.

Environmental Benefits

Conservation agriculture involving zero-till and surface managed crop residue systems are excellent opportunity to eliminate burning of crop residue which contribute to large amounts of greenhouse gases like CO_2, CH_4 and N_2O. Burning of crop residues also contribute to considerable loss of plant nutrients which could be recycled when properly managed. Large scale burning of crop residues is serious health hazard.

(i) **Crop diversification**: Cropping rotations and agro forestry systems when adopted in appropriate spatial and temporal patterns can further enhance natural ecological processes. Indicate that variety of crops such as mustard, chickpea, pigeon pea and sugarcane *etc.*, could be well adapted to the new systems.

(ii) **Resource improvement**: No tillage when combined with surface management of crop residues begins the processes whereby slow decomposition of residues results in soil structural improvement and increased recycling and availability of plant nutrients. Surface residues acting as mulch, moderate soil temperatures, reduce evaporation, and improve biological activity.

Challenges in Conservation Agriculture

Conservation agriculture for raising crops will require an innovative system perspective to deal with diverse flexible and context specific needs of technologies and their management. Conservation agriculture R&D (Research and Development) thus calls for several innovative features to address the challenge. Some of these are:

(a) **Understanding the system**: Conservation agriculture systems are much more complex than conventional systems. Managing these systems understanding basic processes and component interactions which determine the completely performance. For example surface maintained crop residues act as mulch and therefore reduce soil water losses through evaporation and maintain a moderate soil temperature regime (Gupta and Jat, 2010). No-tillage systems influence depth of penetration and distribution of the root system which influence water and nutrient uptake and mineral cycling. Thus need to recognize conservation agriculture system.

(b) **Building a system and farming system perspective**: A system perspective is built working in partnership with farmers. A core group of scientists, farmers, extension workers and other stakeholders working in partnership mode will therefore be critical in developing and promoting new technologies.

Technological challenges: Basic principles of conservation agriculture practices that is no tillage and surface managed crop residues as well as understood adoption of these practices under varying farming situations is key challenge. These challenges related to development standardization and adoption of farm machinery for seeding with minimum soil disturbance developing crop harvesting and management systems.

(d) **Site specificity**: Adapting strategies for conservation agriculture systems will be highly site specific will be a powerful way in understanding why certain technologies or practices are effective in a set of situations and not effective in another set. This learning process accelerates building a knowledge base for sustainable resource management.

(e) **Long-term research perspective**: Conservation agriculture practices *e.g.* no-tillage and surface maintained crop residues result in resource improvement only gradually and benefits come about only with time. Indeed in many situations benefits in terms of yield increase may not come in the early years of evaluating the impact of conservation agriculture practices.

Benefits of Conservation Agriculture

The benefits of the conservation agriculture can be seen at farm regional and national level. The benefits can be classified into three broad categories:

(i) Agronomic benefits that improve soil productivity.

(ii) Economic benefits that improve the production efficiency and profitability.

(iii) Environmental and social benefits that protect the soil and make agriculture more sustainable.

Some of the benefits of conservation agriculture are listed below:

1. Improve the sustainability of different production systems.

2. Provides soil as sink for carbon dioxide by improves soil organic carbon content and contribute in reducing global warming.

   Total potential for soil carbon sequestration by agriculture could reduce about 40 per cent increase in CO_2 emissions (FAO, 2009).
3. Improves water infiltration thereby reduces run-off of surface and groundwater and enhance groundwater recharge.
4. Improves habitation of organisms from larger insects down to soil borne fungi and bacteria which improve soil biological, physical and chemical properties, thereby contribute in increasing crop productivity.
5. Reduce cost of production (15-16 per cent) by saving energy, labour and water, thereby increase farm income.
6. Enhance biodiversity and improves the value of environmental.
7. Reduction in poverty and enhance food and nutritional security due to higher, more stable yields and lower food prices.
8. Greater rural incomes, which are leading to check in rural-urban migration. Yield advantage of zero tillage to rice and wheat by 10-17 per cent over conventional tillage.

Resource-Conserving Technologies

Resource-conserving technologies (RCT) have been developed in order to: reduce the use of and damage to natural resources through agricultural production and increase the efficiency of resource utilization. Most of these technologies target the two most crucial natural resources: water and soil but some also affect the efficiency of other production resources and inputs (*e.g.* labour, farm power and fertilizer). Some of the more popular RCTs, particularly in irrigated or rice-based cropping systems, are described below. Laser levelling for surface irrigated areas it is essential to have a properly levelled surface with the appropriate inclination for irrigation method adopted. Traditional farmer's methods for levelling by eyesight are not sufficiently accurate resulting in extended irrigation times unnecessary water consumption and inefficient water use. Laser levelling reduces the unevenness of field to about ±2 cm resulting in better water application and distribution efficiency improved water productivity increased fertilizer efficiency and reduced weed pressure. Savings of up to 50 per cent in wheat and 68 per cent in rice (Jat *et al.*, 2005).

Bed Planting

Crop is grown on beds and the irrigation water is applied in furrows between the beds. This is common practice for row crops but not for small grain crops such as wheat and rice. The technique advantages such as improved fertilizer efficiency, better weed control and reduced seed rate. It also saves irrigation water compared to a flat inundated field the evaporation surface is reduced and water application and distribution efficiency are increased. Water savings (compared to flat surfaces) reach 26 per cent for wheat and 42 per cent for transplanted rice and yield increases 6.4 per cent for wheat and 6.2 per cent for rice (RWC-CIMMYT, 2003).

Direct Seeding of Rice

Compared with transplanting – may be considered an RCT:

- ☆ It saves labour and fuel.
- ☆ Seeding into dry soil saves water as there is no puddling.
- ☆ Total growing period from seed to seed is reduced by about 10 days.
- ☆ Yields and water efficiency of subsequent rotation crops are increased. On other hand weed management is more difficult in dry direct-seeded rice than in puddled and transplanted rice.

Reduced tillage, zero tillage Intensive soil tillage is the main cause of reduced soil organic matter and hence of soil degradation. Tillage accelerates the mineralization of organic matter and destroys the habitat of soil life. Under zero tillage the mineralization of soil organic matter can be reduced to levels inferior to the input converting soil into a carbon sink (Reicosky, 2001).

Zero Tillage also in Water Saving and Improved Water-Use Efficiency

Water potential saving by zero tillage according to cropping system and climatic conditions. On average water savings of 15 to 20 per cent. However zero tillage can lead to problems with weed control compaction or surface crusting depending on the soil type.

Mulching and Green Manure

The supply of organic matter to the soil through mulching and green manure is important for maintaining and enhancing soil fertility. Mulching material come from crop residues or green manure crops it provides feed for the soil life and mineral nutrients for the plants. If legume crops are used as green manure they can supply up to 200 kg/ha of nitrogen to the soil, In the case of rice fertilizer savings of 50 to 75 per cent. The spreading of mulch on the soil surface reduces evaporation, saves water, protects from wind and water erosion, and suppresses weed growth.

Controlled Traffic Farming

Controlled traffic farming restricts any traffic in the field to same tracks. While these tracks become heavily compacted to the rooting zone does not all resulting in better soil structure and higher yields. The area lost in the traffic zones is easily compensated by better growth of plants adjacent to the tracks so that overall yields are usually higher than in conventional systems with random traffic (Kerr, 2001). Controlled traffic farming is ideal complement to zero tillage or to bed planting systems but it also provides conventional agriculture through time and fuel savings. In the latter case either GPS (global positioning system) guidance or visible bed and furrow systems must be used to limit the tillage operation to the rooting zones and to not disturb the tracks.

Synergies between Resource-Conserving Technologies

Resource-conserving technologies provide scope for synergy example between bed planting and controlled traffic or mulching and zero tillage. Used in isolation any technologies may face specific problems (*e.g.* surface crusting or weeds in direct seeding rice) or have limitations (*e.g.* zero tillage under irrigated conditions). The combination of resource-conserving technologies working in synergy is commonly referred to as "conservation agriculture" (CA). Conservation agriculture Conservation agriculture (CA) is a concept for resource saving agricultural crop production those profits with high and sustained production levels while conserving is environment. CA is based on enhancing natural biological processes above and below the ground.

CA is characterized by three interlinked principles:

- **(i)** Minimum mechanical soil disturbance throughout the entire crop rotation;
- **(ii)** Permanent organic soil cover.
- **(iii)** Diversified crop rotations in annual crops or plant associations in case of perennial crops.

CA adapts to different crops and cropping systems including vegetables, root crops and paddy rice.

Zero Tillage

When the soil is not tilled soil structure is changes. Continuous macrospores are established facilitating water infiltration and soil aeration as well as root penetration into deeper zones. Soil organic matter content increases with higher values near the surface gradually decreasing with depth. Soil macro- and micro fauna and flora are re-established in better soil fertility.

Soil Cover

The permanent soil cover crops such as mulch or green manure cover crops complements the effects of zero tillage by supplying soil organic matter build up soil life which is facilitated by not disturbing the soil. Protecting soil surface the mulch reduces evaporation avoids crusting and suppresses weed growth. Problems experienced in direct seeding or zero tillage are reduced. The application of zero tillage and direct seeding facilitates the management of residues which in conventional systems are often considered a problem.

Crop Rotation

The phytosanitary and weed management benefits crop rotation serves to open different soil horizons with different rooting types. While conventional agriculture "cultivates the land", using science and technology to dominate nature conservation agriculture tries to "least interfere" with natural processes. Surface irrigation applied bed planting in water saving. Under CA the beds are converted into permanent beds and soil tillage is limited to a periodic cleaning and reshaping of furrows. The same permanent bed system is applicable under conservation agriculture for crop rotations which include crops grown on beds for drainage purposes. Permanent bed system leads to controlled traffic another RCT.

Direct Seeding

Direct seeding of crops including paddy rice is possible under zero tillage direct seeding is preferable. Direct seeding in less soil movement than transplanting which often involves some sort of strip tillage. CA facilitates direct seeding by reducing a number of problems encountered direct seeding is applied in isolation (*e.g.* surface crusting or weed control).

Laser Levelling

The benefits of laser levelling in conservation agriculture for conventional agriculture under surface irrigation conditions. To begin with, significant soil movement is required, and so laser levelling is considered an initial investment before converting to a permanent zero tillage cropping system (*i.e.* conservation agriculture). The investment in laser levelling lasts much longer than in conventional systems since under CA no further soil tillage is applied.

Effects of CA

The soil under CA "grows" an average rate of 1 mm per year due to the accumulation of soil organic matter. The organic matter levels rise by 0.1–0.2 per cent per year due to residues left on the soil surface remaining root biomass and reduced mineralization. Within a crop rotation different root systems structure different soil horizons and improve the efficiency of the soil nutrient use. The increased amount of continuous vertical macrospores facilitates the infiltration of rainwater into ground and help recharge the aquifer. The increased soil organic matter levels improve the level of water accessibility to plants 1 per cent of organic matter in the soil profile can store water at a rate of 150 m^3/ha.

The permanent soil cover and the avoidance of mechanical soil tillage reduce the unproductive evaporation of water-use efficiency is increased and a crops water requirements can be reduced by about 30 per cent under both irrigation and rain-fed conditions. The quantitative benefits reduced leaching of soil nutrients and farm chemicals together with reduced soil erosion lead to a significant improvement in the water quality in watersheds.

CA can reduce the overall requirement for farm power and energy for field production by up to 60 per cent compared to conventional farming.

The same is true for mineral fertilizer:

Less fertilizer is lost through leaching and erosion the different rooting systems recycle more soil nutrients from a larger soil volume resulting in improved overall efficiency of fertilizer use in long term to maintain the production and soil nutrient levels over the crop rotation (Saturnino and Landers, 2002).

REFERENCES

Aulakh, K.S. 2005. *Punjab Agricultural University, accomplishments and challenges.* Ludhiana. 23 p.

Barah, B. 2009. Economic and Ecological Benefits of System of Rice Intensification (SRI) in Tamil Nadu. *Agric. Econ. Res. Rev.* 22: 209-214.

Dumanski, J., Peiretti, R., Benetis, J., McGarry, D., and Pieri. C., 2006. The paradigm of conservation tillage, *Proceedings of World Association of Soil and Water Conservation, P1*, 58-64.

FAO. 2002. *Agriculture: towards 2015/2030.* Rome, 420 p.

FAO. 2005. *The state of food and agriculture 2005,* Rome: 197 p.

FAO. 2006. *World agriculture: towards 2030/2050,* Interim report, Rome.

Ghosh, P. K., Das, A., Saha, R., Kharkrang, E., Tripathy, A. K., Munda, G. C., and Ngachan, S.V. 2010. Conservation agriculture towards achieving food security in north east India. *Current Science,* 99(7): 915-921.

Jat, M. L., Malik, R.K., Saharawat, Y.S., Gupta, R. and Paroda, R. 2012. Proceedings of Regional Dialogue on Conservation Agricultural in South Asia, New Delhi, India, APAARI, CIMMYT, ICAR, p. 32.

Jat, M. L., Singh, S., Rai, H. K., Chhokar, R.S., Sharma, S.K. and Gupta, R.K. 2005. Furrow Irrigated Raised Bed Planting Technique for Diversification of Rice-Wheat System of Indo-Gangetic Plains. *Journal of Japan Association for International Cooperation for Agriculture and Forestry,* 28 (1): 25-42.

Jat, M.L., Chandna, P., Gupta, R., Sharma, S.K. and Gill, M.A., 2006. Laser land levelling: a precursor technology for resource conservation, Rice-Wheat Consortium Technical Bulletin Series No. 7. New Delhi, India, Rice- Wheat Consortium for the Indo-Gangetic Plains. 48 p.

Kassam, A. H., and Friedrich, T. 2009. Perspectives on Nutrient Management in Conservation Agriculture. Invited paper, IV World Congress on Conservation Agriculture, 4-7 February 2009, New Delhi, India.

Kerr, P. 2001. Controlled traffic farming at the farm level. GRDC Research Update, Finley, NSW, Australia.

Lal, R. 2004. Soil carbon sequestration impacts on global climate change and food security. *Science* 304: 1623-1627.

Malik, R. K., Gupta, R. K., Singh, C. M., Yadav, A., Brar, S. S., Thakur, T. C., Singh, S. S., Singh, A. K., Singh, R., and Sinha, R. K. 2005. Accelerating the Adoption of Resource Conservation Technologies in Rice Wheat System of the Indo-Gangetic Plains, Proceedings of Project Workshop, Directorate of Extension Education, Chaudhary Charan Singh Haryana Agricultural University (CCSHAU). Hisar, India: CCSHAU.

PDCSR. 2005. Project directorate for cropping systems research. Annual Report 2004–05. India, Modipuram- Meerut. 143 p.

Reicosky, D.C. 2001. Conservation agriculture: global environmental benefits of soil carbon management.

RWC-CIMMYT. 2005. Agenda Notes. 13th Regional Technical Coordination Committee Meeting. RWC-CIMMYT, Dhaka, Bangladesh.

Sangar, S., Abrol, J. P., and Gupta, R. K., 2005. Conservation Agriculture: Conserving Resources Enhancing Productivity, 19 p. CASA, NASC Complex, New Delhi.

Saturnino, H.M. and Landers, J.N. 2002. The environment and zero tillage. Brasilia, Brazil, APDC-FAO. 144 p.

Sharma, A. R., Jat, M. L., Saharawat, Y.S., Singh, V. P. and Singh, R. 2012 Conservation agriculture for improving productivity and resource-use efficiency: prospects and research needs in Indian context. *Indian Journal of Agronomy*, 57 (IAC Special Issue): 131-140.

Wassmann, R., Jagadish, S.V. K., Sumfleth, K., Pathak, H., Howell, G., Ismail, A., Serraj, R., Redona, E., Singh, R. K. and Heuer, S. 2009. Regional vulnerability of climate change impacts on Asian rice production and scope for adaptation. *Advances in Agronomy, 102*, 91-133.

Wolff, P. and Stein, T. M. 1998. Water efficiency and conservation in agriculture–opportunities and limitations, *Agriculture and Rural Development*. 5(2): 17-20.

Conservation Agriculture Technologies *Pages* **225–236**
Editor: **Rahul Kumar, Abhinandan Singh and Pankaj Kumar Ojha**
Published by: **BIOTECH BOOKS, NEW DELHI**

19 Composting of Rural Organic Residues for Sustainable Agriculture

***Sudip Sarkar*[1], *I. Yimjenjang Longkumer*[2] *and Abhinandan Singh*[3]**

[1]Department of Soil Science, [2]Department of Entomology,
[3]Department of Agronomy,
Dr. Rejendra Prasad Central Agricultural University, PUSA,
Samastipur – 848 125, Bihar

Composting of rural organic residues means bio-chemical process, whereinvarious and mixed groups of microbes breakdown organic residues to a humus like material which has properties like farmyard manure. These compost has low bulk density and stability than the native raw organic material, moreover it decompose slowly and can be back to the soil without much devastation of its high energy value.The process of composting is easy to handle and require low costto operate, which can be used as a sustainable farming tool through use of unwanted organic materials for improving soil fertility. Soil health becomes batter by application of organic compost, as it improves various soil physical, chemical and biological parameters of soil, also reduces pollution, foul small and disease-pest.There are various methods adopted for composting based on type, amount and requirement. Composting has no adverse effect on environment and human health, so it can be consider a suitable method of recycling rural organic residues.

Keywords: *Composting, Sustainable agriculture, Soil.*

Introduction

Rising concerns relation to land degradation, improper inorganic fertilizer application, soil pollution, soil health, soil biodiversity and sanitation have reawakened the global interest in organic recycling applications like composting in various parts of world. The high potentiality of composting to convert farm residues as a farm-output into a farm input, whichmakes it an attractive proposal. This process balances the ratio of carbon to other elements, so avoid temporary immobilization of plant nutrients. One of the important benefits of application compost to the soil is that the nutrients present in it are slowly released to the soil and become available for plants. It dealsnumerous benefits such as improved soil fertility and soil health, therefore better agricultural productivity, improved soil biodiversity, lowered environmentalrisks and pushed towards sustainable agriculture. Despite the fact that the practice is well known to many farmers in various parts of the world, butparticularlyin developing countries farmers find difficulty by not making the best use of organic recycling opportunities available to them, due to severallimitations, that includesinefficient technology, long time period, unskilledlabour, land and investment requirements, and profitability.The old traditional method of composting is to pile organic materials and wait for many months or even a year, thanthe materials would be ready for use. The main benefit of this method is low labour requirements, but also has disadvantages like huge space requires for long duration, leaching of nutrients due to exposure to rainfall, and spreading of disease-pest. These limitations could be minimize by introducing advanced and rapid composting methods.

Principles of Composting and Affecting Factors

Organic Materials

Rural compost is prepared from organic residues like various crop residues (straw, leaves, rice husk, groundnut shell, sugarcane trash and bagasse), water hyacinth and other succulent weeds and animal excreta, the organic material undergo exhaustive decomposition under thermophilic and mesophilicstates in heaps or pits along with sufficient moisture and finally produce a dark brown to black colored humified material in three to five months which has more stability in form, valuable for replenishment of plant nutrients, maintenance of soil organic matter and in improving and maintenance of soil fertility (Alexander, 1977).

Carbon-Nitrogen Ratio

It is consider as most important factor in composting of rural residues. The conversion of rural organic materials into compost is mainly a microbiological process and influenced by the shares of carbonaceous and nitrogenous materials present in the residues ready for composting. Microbes require carbon for growth and nitrogen for protein synthesis. Optimum C/N ratio for efficient composting is varies between 30 to 40 and ratios between 25 to 40 give satisfactory result. In case, organic material has low nitrogen *i.e.* a wide C/N ratio, which leads decreasing of biological activityand would be needed to complete degradation of the carbonaceous material. Immobilized nitrogen is recovered afterthe death of the microorganisms and the some degree of nitrogen is thereforerecovered for lowering the carbon content of the residues. The end of the process is thus delayed and completed compost takes longer period to harvest (Chanyasak and Kubota, 1981). Higher C/N ratios are usuallyoccurred by organic materials containing low nitrogen like as cereal straws, trash of sugarcane, cotton stems and jute stalks. It has been found that vary widely C/N ratio between 30 to 300 could occurs in compost produced from rural residues. When available nitrogen is exhausted, under suitable C/N ratios the action of nitrogen-fixing microorganisms dominates and there is an addition of nitrogen from the atmosphere. Raw materials having C/N ratio less than 30: 1 comprising the proportion of excess nitrogen than the requirements of the microbes. In such cases even though the process of decomposition takes place, nitrogen is lost by volatilization as ammonia or viadenitrification. In case lower C/N ratio, ammonia is produced under suitable situations and can be oxidized to nitrite and nitrate but lower C/N ratio under unfavorable situations leads reduction of ammonia. Around 20 to 40 per cent of nitrogen could be lost as ammonia while preparation of compost from animal excreta and water hyacinth. This suggests that mixing of these portions with the carbonaceous material of residues cannot be done in practice andall the time there are pockets of concentrated nitrogenous material which makes an imbalance in the decomposition process along withreduction of nitrogen. It has been found that simple nitrogenous compounds are more prone to loss and reduction of this loss improves the quality of compost (Beri *et al.*, 1992). Duringcomposting higher C/N residues can be lowered by addition of nitrogen source or through blending with organic materials higher in nitrogen like leguminous plant residues.

Shredding of Materials

Shredding of raw materials acceleratedprocess of composting, as small pieces hasgreater surface area exposure, which increase bacterial activity. Particles size less than 5cm are desirable for composting even though larger sizes can be composted adequately. But, the benefit of shredding is doubtful while considering the additional cost involved (Ajay and Jawaid, 2013).

Blending of Materials

The C/N ratio and moisture content are two keyparameters to be considered while mixing different kinds of material for composting. No need of blending if the C/N ratio of the materials is 25 to 50, even though a desirable range is between 30 and 40. Materials low in nitrogen like as sawdust, cereal straw, paper can be mixed with relatively high nitrogenous materials such as leguminous plants andbiogas slurry to get a nearly optimum ratio between 30 and 40. Likewise, wet and dry substances can be mixed together to get a required proportion of moisture and to escape anaerobic conditions. Addition of 5 to 10 per cent soil with moist materials can lower the moisture content and to absorb ammonia when composting materials has low C/N ratio. In case ofinsufficient dry organic material availability, dry soil can be added to compost. Soil can be added to high C/N ratio organic residues to buffer acidicsituations. Addition of soil also gives a more granular structure to freshly finished organic compost.

Moisture and Aeration

Moisture percentage and aeration status are vital factors in composting. In case of moisture in the compost heap is below forty per cent (W/W), then the decomposition will be aerobic but at slow rate. Aerobic decomposition cantake place at any moisture content between 30 and 100 per cent along with adequate turning and avoid higher moisture content. Moisture content ranging from 40 to 60 per cent has been found optimum for good compost preparation (Willson, 1989; Gaur, 1997). For anaerobic composting, moisture content is notsovital as compare to aerobic composting. Experiment shows that, straw have can be used for the maintenance of 80 to 85 per cent moisture content (Das and Keener, 1997).

Temperature

When the microbes multiply in the composting materials, the heat of exothermic biochemical reactions is reserved because oflarger mass, when the temperature increases above 40°C the mesophilic microbes generally replaced by thermophilic microbes. High temperature is impostant for damage of pathogens and weed seeds. The temperatureinside of the heap rises up to 55 to70°C after then it slowly cools to ambient values. Decomposition rate is fastest in the thermophilic range than compare tomesophilic range. The optimum temperature for organic residues decomposition into CO_2 and water has been recorded to be 60°C but it should not exceed 71RC for long as decomposition rate will be lowered due to the thermal kill of microbes.

Reaction

At starting pH value in compost piles is generally somewhat acidic to neutral (Ajay and Jawaid, 2013; Das and Keener,1997). Lime can be added for neutralizing acidity but nitrogen can be preserved without using limestone. It has been found that, by adding of calcium carbonate to compost leads significant losses of nitrogen as the alkaline conditions suitable for volatilization of ammonia.

Microorganisms Involved

The various microbial population changes during aerobic composting. There is a characteristic pattern that the fungi and acid producing bacteria perform during the mesophilic stage. When the temperature rises above 40°C these microorganisms are replaced by thermophilic bacteria and actinomycetes. Mesophilic bacteria and fungi reoccur along with the temperature fall. Mesophilic bacteriapresent a limited time, but, consume most of the easily degradable carbohydrates and proteins, while, thermophilic bacteria consume lipids and hemicelluloses. *Bacillus sp.* is particularly involved in degradation of proteins and amino-acids. Actinomycetes consume starch and water soluble fractions of residues. Degradation of celluloses and hemicelluloses is relatively slow than lipid, while, lignin is most tough substances to decomposition.

Microbial Inoculants

Microorganisms were identified and formulated, those are actively participated during composting. One tone of organic matter can be decomposewithin 90 days by addition of 300g of decomposer (Singh *et al.*, 2012). In case, C/N ratio is very high or when compost piles are poor in microbial content, application of inoculants can accelerate the process of composting.

Control of Pathogens and Parasites

The control of harmful pathogens and parasites while composting is important, especially in case of sewage sludge or night soil contaminated material to be composted, since these can effect human and animal health.

Nature and Quality of Raw Materials

There are various types of raw materials obtainable in rural areas which can be used for compostpreparation. They can be categorized into five groups - Crop remains and weeds, animal wastes, human wastes, kitchen and forest residues (Table 19.1).The farm waste (such as crop residues) are readily available in seasonally, but large quentitity, while waste from kitchen, animal or human are available all year round. Farm weeds can be dried and use for composting or the green weeds can be composted along with dry organic residues obtain from other sources. Succulent weeds like Water hyacinth growsabundantly in water, which can be harvested and composted into manure. Animal shed wastes are produced in huge quantities in rural areas, that burnt as fuel. These can be used as potential resources for mixing with crops remaining during composting (Hall and Sansoucy, 1981).

Table 19.1: Most Commonly available Materials for Composting in Rural Areas

Crop Residues	*Human Habitation Wastes*
Rice husks	Solid wastes (*e.g.* excreta)
Rice straw	Liquid wastes (*e.g.* urine)
Barley Straw	Garbage
Maize cobs	Kitchen wastes
Maize stalks	Vegetable residues
Millet stalks	Fruit residues
Banana stems	Meat waste
Coconut stem	Fish waste
Castor stem	Forest Residues
Castor stem	Non-edible seeds
Cotton stick	Pine needles
Sugarcane trash	Leaves
Cotton leaves	Sawdust
Jamun leaves	
Coconut leaves	
Coconut husks	
Cassava leaves	
Tamarind leaves	
Animal shed Wastes	
Dung	
Vrine	
Biogas slurry	

Source: Livleen *et al.*, 2016.

Kitchen residues can ues as a good blending fraction while composting dry and high C/N ratio farm waste. Carbon-Nitrogen ratio is one of the most serious factors in effective composting, which influence by carbon and Nitrogencontent in the raw materials. Thus, C/N ratio should be known to farmer for specific raw materials, which will be helpful for blending of different types of raw materials for composting. Generally, crop residues (such as straw, stems) has wide C/N ratio ranging from 50 to 120 whereas the green succulent materials (green manure crops and leguminous crops residues), animal residues, vegetable and kitchen wastes are narrow to medium C/N ratio ranging from 10 to 25. Sawdust and rice husk have very wide ratio ranges from 130 to 500, has very slow decomposition rate. Though, sawdust has good water holding characteristics and can be successfully used in composting at up to ten per cent of the total biomass of raw material to be composted (Table 19.2).

Table 19.2: Nitrogen Content and C/N Ratio of Various Available Compostable Residues

Material	*N per cent*	*C:N*
Wheat straw	0.3-0.5	80-130
Rice straw	0.3-0.5	80-130
Barley straw	0.3-0.4	100-120
Cotton stalks	0.6	70
Maize stalks and leaves	0.8	50-60
Sugarcane trash	0.3-0.4	110-120
Lucerne residues	2.55	19
Green weeds	2.45	13
Water hyacinth	2.38	17.6
Ferns	1.5	25
Fallen leaves	0.5-1.0	40-80
Grass clippings	2.15	20
Neem cake	6.05	4.5
Sesbania sp.	2.83	17.9
Cow dung	2.0	19
Human night soil	4.0-6.0	6-10
Human urine	15-18	0.8
Digested sludge	5.0-6.0	6

Source: Livleen *et al.*, 2016.

Methods of Composting

Indore Method

The materials needed for composting are mixed plant residues, animal dung and urine, earth, waterect. Hardyorganic materials (such as hardy stalks, sawdust, grain husk) should not exceed 10 per cent of the total plant residues. Green materials which are soft and succulent are kept for drying to remove excess moisture before being piled. Make small pieces of materials to increase microbial activity. Each material is extent in layers of about 15 cm thickness up to the pile is about 1.5 m high. The mixture of different kinds of vegetable wastes makes decomposition processmore efficient(Goluekem, 1991). The heap is then cut vertically andaround 20 to 25 kg is put on the floor of cattle shed as bedding material for overnight. The next morning the bedding materialalong with the dung, urine and urine-earth is taken up to the composting pit (BSI, 2005).

A) Pit Method

Site and Size of Pit

The site should be at high level to avoid entry of rain water during the rainy days, it should be near to cattle shed. Compost pit should be about 1 m deep, 1.5 to 2.0 m wide and of any favorable length (Mine Engineer Community, 2000).

Filling Pit

The material taken from the cattle shed is spread and dung slurry made up of 4.5 kg dung, 3.5 kg urine-earth and 4.5 kg of inoculums taken from a 15 days old compost pit. Aadequate quantity of water *i.e.* nearly 90 per cent is sprinkled over the material in the pit to maintain proper moisture content and avoid compacting the material in pit. The pit is complete in this way layer by layer and it should not take more than one week to finish.

B) Heap Method

During the high rainfall the composting can be done in heaps above ground and secure by shed.

Dimensions

The pile is about 2 m wide at the base, 1.5 m high and 2 m long. The sides of pile are tapered to make the top about 0.5 m narrower in width than the base.

Forming the Heap

The Indore heap is generally made with a 20 cm layer of carbonaceous substances such as leaves, straw, stalks, sawdust and wood chips, which will covered with 10 cm of nitrogenous material like as fresh grass, plant residues, garbage, manure, swage. Likewise filling should do until the pile is 1.5 m high. The pile should cover with soil or hay to maintain heat and is turned at six and twelve week intervals. All materials can be mixed together in the pile by carefully maintaining the proper shares of materials.

Advantage and Limitation

Indore method of composting is suggested where land and labor can get easily. Even though, the labor requirements are comparatively high but compost preparation can provide employment in rural areas. There is lack of protection from rain and wind in this method, though it can be upgrade up to an extent. A large amount of water is required and so it is not favorable in areas of low rainfall. The aerobic decomposition leads to huge losses of nitrogen, therefore care should be taken to maintain the C/N ratio about 30 to 40.

Windrow Method

Windrow method of composting for huge amountneeds to create of windrows with a length of 40 m, which comprises of generally paddy straw and leaves in a ratio of 1: 3(volume basis). Every windrow should consist of paddy straw, cow dung, soil, old compost in the ratio of 8: 1: 0.5: 0.5. The biomass constituents along with compost inoculants are mixed carefullywith 3 to 4 days old cow dung before starting windrow. Windrows are kept in trapezoidal formwith a bottom width of 2 m and top width of 1.5 m to match equalwith the width of the rotor of the Turner-cum-Mixer. To increase the process of composting, PUSA Compost Inoculants developed at IARI, which can apply at a rate of 1000 ml per tone of the material.

Machines for Windrow Composting

Various machines are used for window composting namely Compost Turner-cum-Mixer, Compost Sieving machine and a tractor are useful for good quality compost producing and handling. The important parts of the machine are mixing rotor and a water tank to store water and culture. Rotor has blades for turning and mixing the materials.Now a days Automatic Compost Sieving Machine (ACSM) is good for sieving and separating the raw material. A tractor having front mounted loader is suitable for the purpose of handling of materials such as loading, transfering and shaping of the compost materials. The mechanization of the windrow composting makes easy for a large scale production of good quality compost with laborers.

Advantage and Limitation

Windrow method is economically and commercially suitable compost preparation method. Composting by manually is labor intensive, which canovercome by use of machines. Through this method compost get ready in 45-60 days, whereas in traditional pit method it can takes more than 120 to 140 days based on the raw materials and seasonal differences.

The Bangalore Method

This method is mainly recommending while night soil and refuses are used for making the compost. This method can overcome many of limitations of the Indore method but the time period required for composting is much longer. The method isespecially favorable for areas with a low rainfall.

Preparation of the Pit

Pitsare made with a depth of one meter; the width and length of the pit can be prepared based on the land availability and the type of material to be composted. The site selection for the pits is made same as in the Indore method. The trenches should suitably have sloping walls and a floor of 90 cm slope to avoid water logging.

Filling the Pit

Organic waste and night soil or animal dung are placed in alternate layer and after finishing the pit is covered with a thick refuse layer ranging from 15 cm to 20 cm. The raw materials are leavein the pit with no turning or watering for three months. During this period, material settles down due to decline in volume of the biomass and some new layers of night soil and refuse placed in alternate layer and finally covered with soil to prevent loss of moisture, control foul odor and flies. The initial aerobic composting takesaround eight to ten days, after then anaerobic decomposition goes on at a slow rate, which takes about six to eight months to obtain the finished product.

Advantage and Limitation

The recovery of the composted product is relatively higher then aerobic composting and also the nitrogen loss is negligible. There are low labor requirement as compared to Indore method, since, turning of materials is not done. This method requires more time to complete the composting process and hence, it takes up more space. Special care should be taken to control foul smell and flies.

Rural Composting in India

The Digging and Preparation of a Compost Pit

The pits can be divided into two parts with dividers made from hardy straw materials. The sources are usually animaldung along with straw, husks and used as bedding in animal shed and also weeds in addition to crops remaining. These waste materials are spread in the pit layer by layer and superphosphate added.

High Temperature Heap Composting

This type of compost is prepared generally from human residues, animal residues and chopped residues at ratio 1: 4. Heap is preparedthrough alternate layers of chopped residues and human or animal waste, water is according to need. Bamboo poles are inserted into heap for abetter aeration. After the completion of heap, it should be plastered with a mud layer of 3 cm. Within few days the temperature rises to 60-70°C and bamboo poles are removed followed sealing of holes with mud. Composting is done within two to three months.

The Merkley Method

This method of composting was established at the University of California at Berkley in 1953. Compost can be ready within two week through the method.

Raw Materials

A mixture of organic residues in the ratio of 2: 1 of cellulose rich material and nitrogen rich materials should be used for a rapid composting. It is suggestedto that the straws, leaves, grass and manure should be mixed and chopped with a mechanical device before the heap preparation. Papers and hardy materials should not exceed ten per cent of total raw material.

Dimensions and Procedure

The compost heap length, width and height should be 2.4 m, 2.2 m and 1.5 m respectively and composting to be done for two weeks. Turning and mixing is repeated on the fourth, seventh and tenth days. Compost is ready for use from two weeks,but, to get finer compost is required to furtherdecomposeand can be sieved.

Advantages and Constraint

The main benefit of the Berkley method is fast composting of large quantities material, even though, it is labor intensive due to more turning requirement. Materials with a narrow C/N ratio may takes more time and not suitable as raw input for this method.

Nature and Properties of Finished Compost

Organic materials should sufficiently decompose and should have positive effect on the growth of plant. Humus may be defined as a complex aggregate, dark brown to black in color substances resulting from the microbial breakdown of plant and animal wastes under aerobic or anaerobic situation.Humus is heterogeneous material composed of different type of polymerized substances such as aromatic molecule, hemicelluloses, cellulose, lignin and proteins. Generally, finished compost is characterized bydark brown to black in color,insoluble in water, C/N ratio of 10 to 20, highcation exchange capacity, water absorption with rapid swelling, beneficial effects on soil and plants and finally it should enhance fertilizer use efficiency.

Methods for Determining the Condition of Compost

There are many tests used to check the condition of compost at different stages interval. Any single test is not enough to determine the compost condition. Important physical features are temperature, texture, color and solubility in NaOH. C/N ratio and nutrient content of compost is essential factors along with nonappearance of flies and foul odour. Generally, in small scale or garden composting colour, texture, temperature and effect on health are considered, while in large commercial production chemical parameters also taken into concern.

Temperature

It is considered best indicator of the process, indicates composting stage, presence of pathogen and weed seed.In the initial stages of composting process the active microbial population increases exponentially until the substrates are available or other growth limiting factors arises (Marugg *et al.*, 1993). In case of aerobic composting quick rise of temperature to 55RC to 65RC in the first four to five days and this high temperature continue active decomposition in presence of proper moisture. Failure of temperature in this period hampered decomposition process. After heating period a stable cooling point indicates finishing of composting. The checking of temperature is done through a thermometer fitted on the tip of a stick length ranges 0.5-0.75 m. The thermometer should enter the pile for 5-10 minutes in different points and take the mean value for judgment.

Structure

Compost should be crumbly and loose enough to spread in field.

Colour

Blackish to brown color is considered best. Pure black color indicates unfavorable fermentation and presence of excess moisture, while, grayish or yellowish color indicates to much of sol or ash and greenish colour indicates incomplete or partially complete decomposition.

Moisture

Simply insert an iron rod and check the moisture by visual observation. Mositure status can also check through pressing the sample by hand.

Odour

Foul odor indicates incomplete decomposition. By disturbing the heap smell can be detected. Good compost should smell like humus or soil.

Acidity

Neutral to slightly acidity is considered the best even though slight alkalinity is also adequate.Good compost pH ranging 6.0 to 7.5 and nitrogen fixing bacteria are most active at this pH range.

C/N Ratio and Fertilizer Value

The share of raw material and moisture status determines the C/N ratio and fertilizer value of composted material. Generally, compost should have around 30-60 per cent organic matter and C/N ratio as low as 20-25 in finally produced material. Chemical test like NPK content, CEC and heavy metal test can helps to make sure the quality of compost and its fertilizer value.

Future Prospects

New effective approaches need to beintroduced to improve the process of composting, comprising reduction of composting period, addition of more animal and human waste and improve the activities ofbeneficial microorganisms. The composted material should availablefor diverse agriculture and considering an essential component of crop production. Based on climatic, local resources and demand, suitable technology and inoculants should develop to meet the requirement of local farming community. Even though many of work has been already done on microbiology of composting, but still there is inadequate knowledge about active microbial population and their role in various phases of composting, so this research gap should be fulfilled. Unculturable microorganisms can studied through advanced technology like gel electrophoresis.Extension department need to play important role in promoting effective compost technologies among farming community. Finally Govt. and public should work together to extend the recycling of organic materials, which will help to achieve sustainability in current agriculture.

REFERENCES

Ajay, K. andJawaid, S.M. (2013). Soil Modification Using Shredded Scrap Tires. *International Journal of Biological Sciences and Technological Research*, 1: 10-13.

Alexander, M. (1977). *Introduction to Soil Microbiology*. Wiley and Sons Inc New York, USA.

Beri, V.,Sidhu, B. andBhat, A. (1992). *Nutrient Management for Sustained Productivity*, p. 133-135.

BSI PAS (2005). Specification for Composted Materials British Standard Institution, London, UK, 100: 2005.

Chanyasak, V. and Kubota, H. (1981). Carbon/organic nitrogen ratio in water extract as measure of composting degradation. *Journal of Fermentation Technology*, 59(3): 215- 219.

Das, K. and Keener, H.M. (1997). Moisture effect on compaction and permeability in composts. *Journal of Environmental Engineering*, 123(3): 275-281.

Gaur, A.C. (1997).Bulky organic manures and crop residues. In: Tandon H.L.S., Ed., Fertilizers, Organic Manures, Recyclable Wastes and Biofertilizers, FDCO, New Delhi, India, pp. 37-51.

Goluekem, C. (1991).The Staff of Bio Cycle Journal of Waste Recycling. The Art and Science of Composting, The JG Press Inc, Pennsylvania, USA, pp. 14-27.

Hall, J.M. andSansoucy, R. (1981).Open yard housing for young cattle. FAO Animal Production and Health Paper No 16, Rome, Italy.

Livleen, S., Satish, D.L., Indra, M.M., Archna, S. and Annapurna, K. (2016). Rural Composting for Improvement of Soil Health and Sustainable Agriculture. *Agri Res and Tech: Open Access J.*, 1(5): 1-8.

Marugg, C., Grebus, M., Hansen, R.C., Keener, H.M. and Hoitink, H.A.J. (1993).A kinetic model of the yard waste composting process. *Compost Science and Utilization*, pp. 39-51.

Singh, S., Singh, B., Mishra, B.K., Pandey, A.K. and Nain, L. (2012).Microbes in Agrowaste Management for Sustainable Agriculture.Microorganisms in Sustainable Agriculture and Biotechnology Springer, Netherlands, pp. 127-151.

Willson, G.B. (1989).Combining raw materials for composting. *Biocycle*, 82-85.

Conservation Agriculture Technologies *Pages* **237–246**
Editor: **Rahul Kumar, Abhinandan Singh and Pankaj Kumar Ojha**
Published by: **BIOTECH BOOKS, NEW DELHI**

20 Biopesticides: A Radical Shift from Chemical Pesticides

I. Yimjenjang Longkumer[1], Sudip Sarkar[2] and Abhinandan Singh[3]

[1]Department of Entomology,
[2]Department of Soil Science and Agricultural Chemistry,
[3]Department of Agronomy,
Dr. Rajendra Prasad Central Agricultural University, PUSA,
Samastipur – 848 125, Bihar

Biopesticides is a contraction of biological and pesticides which are ecofriendly pesticides and obtained fromnaturally occurring substances, microbes and plants. They tend to have low risk to the nontarget organism and also do not harm the ecological and environmental process. Some biopesticides agents when applied once can produce pest control from the subsequent generation also. On the contrary to the hazardous impact caused by the chemical insecticide which deliberately hampers the ecology of natural enemies, nutrient cycles etc. India has a prodigious potential for biopesticides. Theyare highly host specific and are surmised to be relatively safe to non-target organism including humans. They should be encouraged as an alternative to chemical pesticides and their potential in agricultural pest management cannot be ignored. Through this review, I would relish to highlight the role of biopesticides as a radical shift in managing pest in various agroecosystem.

Keywords: Biopesticides, Environment, Chemicals, Potential.

Introduction

Inthiscontemporary world, in agricultural pest management perspective, the farmers are highly onto and dependent on the chemicals for the management practices. With the fact of providing instant action, highly effective *let alone* the good yield it produce. It has been estimated that about 2.5 million tons of pesticides are used on crops each year and the worldwide damage caused by pesticides reaches

$100 billion annually (Koul *et al.*, 2008). Application of pesticide brings about many negative impacts in the environment. Over 98 per cent of sprayed insecticides and 95 per cent herbicides fall on the non-target siteother than their target species living in various ecosystems. The main reason leading to water pollution is the application of pesticides; some are persistent organic pollutants contributing to soil contamination (Bradman, 1999). Due to its misuse and overuse, residues that are present in foods, fruits, vegetables and even in mother's milk creates an alarm on human health (WHO, 2002). Therefore, initiation of alternative plant-based pesticides approach could be a possible and viable solution to counteract this catastrophe which could serve as potent "Ecofriendly-agrochemicals" (Govil and Sanjib, 2012).One insecticidal class belonging to organochlorine compoundscould virtually affect the every form of life in this earth (Hurley *et al.*, 1998). Some chemicals like endocrine disruptor pesticides (*e.g.* DDT) elicit their adverse effects by acting on the body system mimicking and antagonising the functions of the hormones and these has been reported to have several negative impact on the humans such as immune suppression, hormone disruption, reproductive abnormalities and cancer (Brouwer *et al.*, 1999; Crisp *et al.*, 1998; Hurley *et al.*, 1998).Soil degradation is also a common phenomenon which arises due to high usage of pesticides. This hampers the nutrient cycle in which plants require for their normal growth and development. The techniques employed for the alternate approach to the pesticides are empowering and will eventually lead to the reduction of over usage of chemical insecticides. However, the other disastrous sides of employing chemicals are not clearly understood by them as they are already captivated with the efficient result provided from the chemicals. The most progressive and effective approach as an alternative to pesticides is the implementation of naturally occurring substance or living organism termed as Biopesticides. It involves the usage of naturally occurring plant products, natural enemies and microorganism. With this approach the negative impact of pesticides can be avoided which will eventually lead to a sustainable and ecofriendly environment maintaining and protecting all the reliable resources which balances the life form. Through this chapter, I would relish to highlight the role of biopesticides as a radical shift from chemical pesticides in managing pest in various agroecosystem.

Significance and Impact of Biopesticides

The term biopesticides envisage those compounds which are naturally present and have specific host in managing on contrast with broad range host in chemical pesticides. It refers to products containing biocontrol agents *viz.*fungi, bacteria, *etc.* or compounds derived from naturally occurring materials (plantsor certain minerals), including their genes or metabolites, for controlling pests. According to the FAO "Biopesticides arepassive biocontrol agents, in contrast to biocontrol such as parasitoids, predators, and many species of entomopathogenic nematodes which actively feeds on the host". Biopesticides have high potential products and can be classified as:

1. Microbial Pesticides

It includes pesticides that contain fungi, bacteria, nematodes virus *etc.* as an active ingredient. They have high host specificity, efficient tool for managing the pest, when once applied can provide from the subsequent generation also.

2. Plant-Incorporated Protectants (PIPs)

It involves the introduction of desired genes in to the plant system upon which the pest once feed, the plants will produce the effect due to the genes genetically incorporated inside its system. (*e.g.*, *Bt* cotton).

3. Biochemical Pesticides

It includes naturally occurring substances with no toxic property and efficiently provide good result in managing pests' problems. Such as botanicals, plants extracts, and pheromones.

Biopesticides are inherently harmful to the non-target organism thanchemical pesticides. They have very high host specificity towards its applicationhaving low or considerable risk to the environment. A minute quantity of its application can provide an efficient result. Hence it can be employed as a supplementary system towards pest management (Ramarethinam, 2002). Among the biopesticide, the neem seed kernel extract have feeding, oviposition deterrent, inhibitors *etc.* biochemical property (Natarajan and Sundaramurthy, 1990). *Bacillus thuringiensis* (*Bt*) is more commonly used against lepidopteran larvae.

Entomopathogenic Fungi (EPF)

Entomopathogenic fungi are natural regulators of pest populations and have high potential as mycoinsecticide agents against diverse insect pests in agriculture. They infect their host by penetration of their cuticle in to the host which is followed by entrance to the haemolymph and release of toxins (Hajek and St. Leger, 1994). They are applied in the form of conidia or mycelium in which after spraying they sporulate. They can also be effectively employed as an alternative to chemical insecticides in insecticide resistance management (Hoy, 1999). Insect pest are known to develop resistance easily against chemicals hence utilization of this agents is gaining interest in pest management worldwide. Steinhaus (1958) reported the compatibility of pathogens and insecticides as one of the advantages of microbial pest control. Franz (1961) provided an excellent review on several fungal entomopathogens in combination with sublethal doses of insecticides against important insect pests. Entomopathogenic fungi mostly belong to Zygomycotina, Ascomycotina, Basidiomycotina and Deuteromycotina. There are around 88 genera of knownentomopathogenic fungi of which only 6 are commercially available. Due to several mode of actions shown by the EPF the pest do not developed resistance against them. Recently, multiple or single foliar applications of *Beauveria bassiana* strain GHA (Bb-GHA) have been found effective for the management of Colorado potato beetle *Leptinotarsa decemlineata* (Wraight and Ramos, 2015). Albeit it's high potential in managing pest there lies some constraints such that exposure to abiotic stress hampers its efficiency. The Table 20.1 will highlight the most promising EPF that are utilized in pest management.

Table 20.1: Examples of Promising EPF and their Target Host

EPF	*Target Pest*	*Dosage*
Metarhizium anisopliae	White grub, sweet potato weevil, BPH, DBM, Rhinoceros beetle, termite, grass hoppers, caterpillars	1-2 kg/ac spray,10-15 kg with 50 kg FYM/vermicompost soilapplication
Beauveria bassiana	Leaf hoppers, plant hoppers, whiteflies, caterpillars and DBM	0.4-1.0 kg/ac Foliar spray
Verticillium lecanii	Coffee green scale, Leaf and plant hoppers, coffee berry borer *etc.*	0.4-1.0 kg/ac Foliar spray
Nomuraea rileyi	Leaf eating Caterpillars, *S. litura, H. armigera,M. separata, A. ipsilon etc.*	0.4-1.0 kg/ac Foliar spray
Hirsutella thompsani	Coconut Mites	1-5 g/l of water

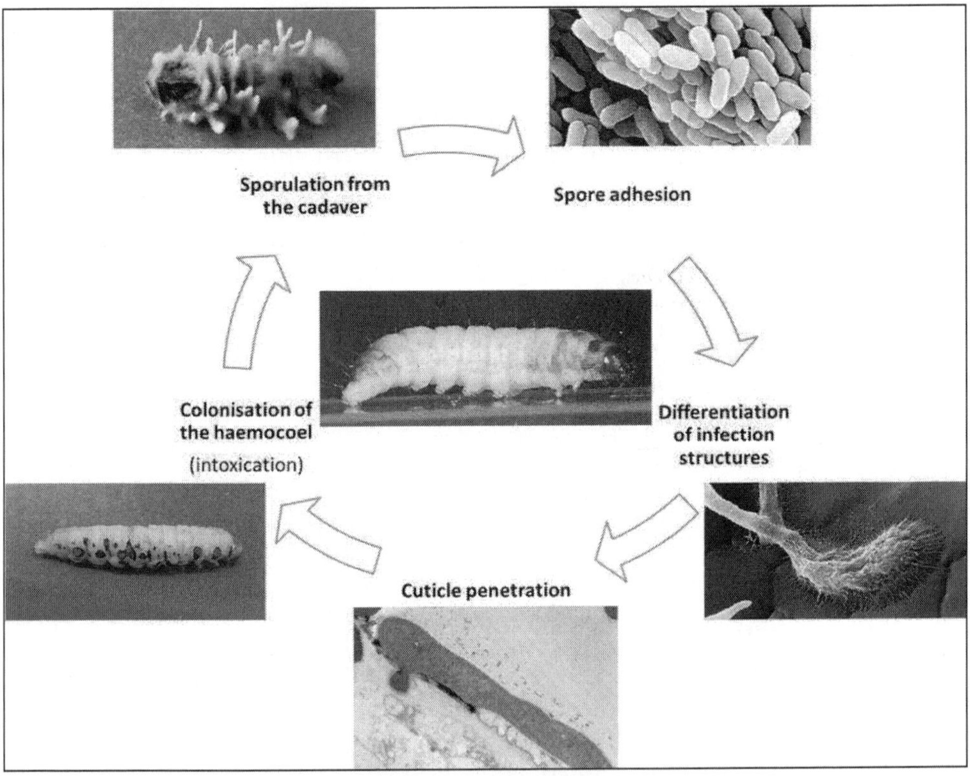

Figure 20.1: General Overview of the Invasive and Developmental Processes of Entomopathogenic Fungi in an Insect Host (Butt *et al.*, 2016).

Entomopathogenic Bacteria (EPB)

The paradigm of pest control by application of biopesticide arose a changed some decades later, when the potential of entomopathogenic bacteria was discovered, of which the species belonging to the genus *Bacillus*were reported (Glare and O'Callagan, 2000). Initially, the species *Paenibacilluspopilliae* was introduced for the control ofthe Japanese Beetle *Popillia japonica* Newman (Steinhaus, 1975), but more effective result were developed when *Bacillusthuringiensis* were discovered at competitive level with the conventional insecticides in terms of efficacy and production cost. The strain HD-1, belonging to subsp. *kurstaki* (De Barjac and Lemille, 1970), became the most effective strain for the management of lepidopteran pests. After few decades other strains are commercially available, such as SA-12, PB 54, ABTS- 351 and EG2348, which are isoloated from the soil, insects *etc.* and possess several toxins from the cry1 and cry2 families. Subsequently, it was followedwith the discovery of a Bt subsp. *israelensis* (Bti) for the management of mosquitoes and simulids (Goldberg and Margalit, 1977). An active strain was also developed *i.e.* Bt subsp. *tenebrionis*which proved to be very effective against the coleopterans (Krieg *et al.*, 1983). Thus, BT became a cornerstone as an effective tool in pest management. The mode of action involves the production of parasporal bodies (crystals) containing insecticidal endotoxins (Cry proteins) through which upon ingestion by the pest a pore-forming mechanism detrimental against insect gut epithelium occurs (Pigott and Ellar, 2007). Due to strain-specific mode of action, cry toxins act by binding to gut receptors which vary among insect species. Hence, an increasing number of cry genes have been identified and sequenced, and more genes continue to be discovered as new bacterial isolates are collected worldwide (De Maagd *et al.*, 2003). The Table 20.2 shows the most promising EPB employed in the management of pest in agroecosystem.

Table 20.2: Promising Bt Strain and its Target Host

Bt subsp. Strain	*Trade Name*	*Target Pest*
kurstaki	Able, Bactospeina, Condor, Costar, CRYMAX, Cutlass, Dipel ES, Bactimos L, Futura, Lepinox, Thuricide	Lepidoptera
aizawai	Florbac, Agree, Design, Xentari	Lepidoptera
kurstaki SA-12	Costar	Lepidoptera
kurstaki	Foil, Raven	Lepidoptera/Coleoptera
kurstaki HD-1	Thuricide, Biobit, Dipel, Foray, Javelin, Vaul	Lepidoptera

Kalha *et al.*, 2014

Entomopathogenic Viruses (EPV)

Entomopathogenic viruses are obligate intra-cellular parasites having either DNA or RNA as their genome which are encapsulated into a protein coat (cap-sid) forming the virions (infective particles). Their efficacies have caught the attention of many researchers and are employed for the management of pestsbelonging to order Lepidoptera and Hymenoptera worldwide. Like EPB, they have high host specificity the diseases caused by EPV have been known since the 16thcentury. Nucleo polyhedrosis virus was observedin 1524 by Vida in a silkworm rearing facilities which caused a disease namely *jaundice grasserie.* In 1856, two Italian scientists, Maestri and Cornalia, first described the occlusion bodies (OBs) of silk-worms. In general, viruses are divided into two broad categories:

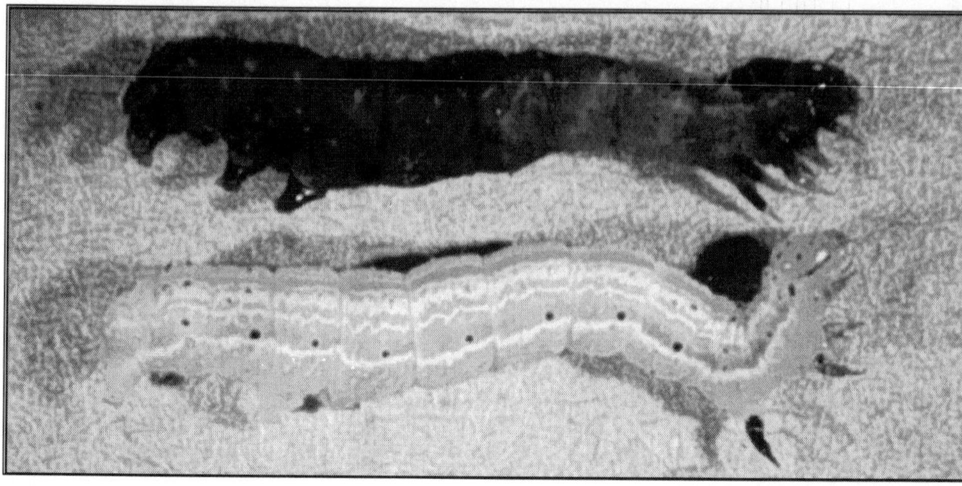

Figure 20.2: Caterpillar Infected by *Bt* (up), Fresh Caterpillar Not Infected by *Bt* (Down) (Mahr, 2004).

1. Occluded-viruses and
2. Non-occluded viruses.

Occluded-viruses are characterized by the presence of distinct mature virion particles (virions) which are embedded within a protein matrix, forming para-crystalline bod-ies that are generally referred to as OBs.

Non-occluded viruses are free virions that form para-crystalline bodies, characterized by the absence of occlusion protein that is inter-spersed among the virions (Federici, 1999).

Out of a total of 73 known virus families, EPV have been listed in 13 families as described by Murphy *et al.* (1995). Among these 13, Baculoviridae fam-ily members are the most virulent on different orders of insect-pest including Lepidoptera, Diptera, Hymenoptera, Orthoptera, Isoptera and Neuroptera. Currently, Baculoviridae is divided into two genera: *Nucleopolyhedrovirus* (NPV) and *Granulovirus* (GV) (Francki *et al.*, 1991). Virions of NPV and GV are occluded in polyhedral and capsu-lar proteinaceous OBs, respectively. The size of OBs of GVs ranges from 0.3–0.5 µm in length than those of the NPVs (0.15–15 µm in diameter) and contain only a single enveloped nucle-ocapsid. The OBs of NPVs contain several hun-dred virus particles, each of which may contain one or many nucleocapsids. NPVs have narrow host ranges, specific to either one host species or genus, with the exception of the NPVs of *Autographa cali-fornica* (Speyer), *Anagrapha falcifera* (Kirby) and *Mamestra brassicae* (Linnaeus). On the other hand GVs are more host specific than NPVs and they have been only reported from Lepidoptera (Moscardi, 1999). Baculoviruses have a large, double-stranded, circular DNA genome of size 88 and 200 kbp. The baculoviruses are characterized by the presence of rod-shaped nucleocapsids, surrounded by a lipoprotein envelope to form virus particles. NPVs pro-duce large particles within the nucleus of an infected cell. The occlusion body is composed of a matrix comprising a 29 kDa protein known as polyhedron. The

DNA–protein complex is contained by a rod-shaped nucleocapsid com-prising a 39 kDa or 87 kDa capsid protein (King *et al.*, 1994). On the other hand, GVs con-tain one virion (singly enveloped nucleocapsid) per virus occlusion body or granule. Granulin, the major granule protein, is similar to poly-hedrin in function. The baculovirus life cycle involves two distinct forms of virus:

1. Occlusion/Polyhedra-Derived virus (ODV/PDV) is present in a protein matrix(polyhedrin or granulin) and is responsiblefor the primary infection in the midgutepithelial cells of the host.
2. Budded virus (BV) is the non-occluded formreleased from the infected host cells laterduring the secondary infection.

Table 20.3: Entomopathogenic viruses that have been used for biological control of insect pests

Common and species names	*Targeted insects*
Corn earworm NPV (HezeSNPV)	*Helicoverpa zea, Heliothis virescens*
Cotton bollworm NPV (HearNPV)	*Helicoverpa armigera*
Diamond back moth GV (PlxyGV)	*Plutella xylostella*
Unbarred Spodoptera moth (army wormNPV (SdalNPV)	*Spodoptera albula*
Beet armyworm NPV (SpexMNPV)	*Spodoptera exigua*
Tobacco armyworm NPV (SpltNPV)	*Spodoptera litura*
Gypsy moth, NPV (LydiMNPV)	*Lymantria dispar*
Potato tubermoth GV (PhopGV)	*Phthorimaea operculella*
Cabbage looper (TrniSNPV)1	*Trichoplusiani*

Lacey *et al.*, 2015.

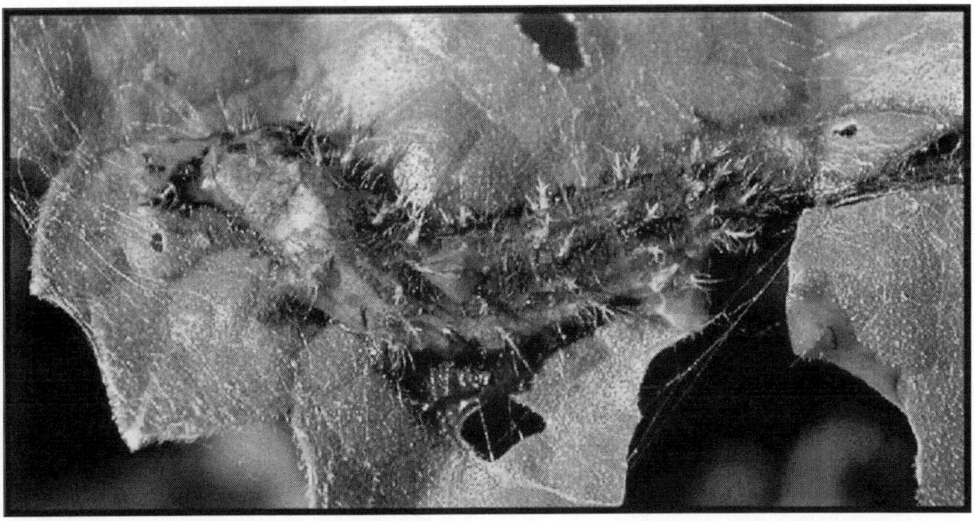

Figure 20.3: Example of Lepidopteran Larva (Thistle Caterpillar) Infected with Nuclear Polyhedrosis Virus (Knodel, 2008).

Conclusion

The role of biopesticides as an alternative to chemical pesticides and its efficient result has caught a high attention to the researchers. Its application in organic farming and production of pesticide residuefree agricultural produce would certainly convince the farmers for its adoption. Effective Training on dissemination of its production and quality control in manufacturing and organizational training to extension workers and farmers to broaden the scopeof biopesticides may invoke an essential channel for implantation of biopesticides. Globally, more emphasized is given against environmental safety as far as pest management is concerned, thus given the safe side of the biopesticides, it would strengthen more when the end users are made aware of the hazardous impact caused by the chemical pesticides, these will ultimately evoke a radical change of biopesticides from chemical pesticides. It is also believed that biopesticides causes less vulnerable to genetic variations in plant populations that cause problems related to pesticide resistance. If deployed appropriately, biopesticides have potential to enhance sustainability to global agriculture for food and feed security.

REFERENCES

Bradman, H. (1999). The Global Distribution of Fatal Self Poisoning: Systematic Review. *BMC Pub Health,* 7: 5-7.

Brouwer, A., Longnecker, M. P., Birnbaum, L. S., Cogliano, J., Kostyniak, P., Moore, J., Schantz, S. and Winneke, G. (1999). Characterization of potential endocrine related health effects at lowdose levels of exposure to PCBs. *Environ Health Perspect,* 107: 639.

Butt, M. T., Christopher, J. C., Dubovskiy, I and Ratcliffe, N. A. (2016). Entomopathogenic Fungi: New Insights into Host-Pathogen Interactions. *Advances in Genetics* 94: 307-364.

Crisp, T. M., Clegg, E. D., Cooper, R. L., Wood, W. P., Anderson, D. G., Baeteke, K. P., Hoffmann, J. L., Morrow, M. S., Rodier, D. J., Schaeffer, J. E., Touart, L. W., Zeeman, M. G. and Patel, Y. M. (1998). Environmental endocrine disruption: An effects assessment and analysis. *Environ Health Perspect,* 106: 11.

De Barjac, H. and Lemille F. (1970). Presence of flagellar antigenicsubfactors in serotype 3 of *Bacillus thuringiensis. Journal of Invertebrate Pathology,* 15: 139-140.

De Maagd, R. A., Bravo, A., Berry, C., Crickmore, N. and Schnepf, H. E. (2003). Structure, diversity, and evolution of protein toxins from spore forming entomopathogenic bacteria. *Annual Review of Genetics,* 37: 409-433.

Federici, B. A. (1999). A perspective on pathogens as biologicalcontrol agents for insect pests. In: Handbook of Biological Control: Principles and Applications of Biological Control(T. S. Bellows and T. W. Fischer, Eds.), Academic Press, CA, pp. 517–548.

Francki, R.I. B., Fauquet, C.M., Knudson, D. L. and Brown, F. (1991). Classification and nomenclature of viruses: Fifth Report of the International Committee on Taxonomy of Viruses. Arch. Virol. 2, 117-123.

Franz, J.M. (1961). Biological control of pest insects in Europe. Ann. Rev. Ent., 6: 183-200.

Glare, T. R. and O'callaghan, M. (2000). *Bacillus thuringiensis*: Biology, ecology and safety. Wiley, Chichester, UK.

Goldberg, L. J. and Margalit, J.(1977). A bacterial spore demonstrating rapid larvicidal activity against *Anopheles sergentii, Uranotaenia unguiculata, Culex univittatus, Aedesaegypti and Culex pipiens*. Mosquito News, 37: 355-358.

Govil, J. N. and Sanjib, B. (2012). Neem Oil: biological activities and usage. "Recent Progress in Medicinal Plants" (RPMP)" Fixed Oils and Fats of Pharmaceutical Importance, Vol 33, Stadium Press.

Hajeck, A. E. and Leger, R. (1994). Interactions between fungal pathogens and insect hosts. *Ann. Rev. Ent.*, 39: 293 - 322.

Hoy, M. A. (1999). Myths, models and mitigation ofresistance to pesticides. In: Insecticide Resistance: From Mechanisms to Management (I. Denholm, J. A Pickett and A. I. Devonshire Eds.), New York, CABI Publishing, pp. 111-119.

Hurley, P. M., Hill, R. N. and Whiting, R. J. (1998): Mode of carcinogenic action of pesticides inducing thyroid follicular cell tumours in rodents. *Environ Health Perspect* 106: 437.

Kalhal, C. S., Singh, P. P., Kang, S. S., Hunjan, M. S. and Sharma, R. (2014). Entomopathogenic viruses and bacteria for insect-pest control. (Rishu Sharma eds). p. 225.

King, L.A., Possee, R.D., Hughes, D.S., Atkinson, A.E.,Palmer, C. P. and Marlow, S. A. (1994). Advances in insect virology. *Adv. Insect Physiol.* 25, 1–73.

Knodel, J. J. (2008). Bertha Armyworm in Canola: Biology and Integrated Pest Management. www.ag.ndsu.edu.

Koul, O, Walia, S. and Dhaliwal, G. S. (2008). Essential Oils as Green Pesticides: Potential and Constraints. *Biopestic Int* 4: 63-84.

Krieg, A., Huger, A. M., Langenbruch, G. A. and Schnetter, W. (1983).*Bacillus thuringiensis* var. *tenebrionis*: a new pathotype effective against larvae of Coleoptera. *Journal of Applied Entomology*, 96: 500-508.

Lacey, L. A., Grzywacz, D., ShapiroIlan, D. I., Frutos, R., Brownbridge, M. and Goettel, M. S. (2015). Insect pathogens as biological control agents: Back to the future. *Journal of Invertebrate Pathology*, 132 (2015) 1–41.

Mahr, S. (2004). General Approaches to Insect Control. Available online at https: // wimastergardener.org/article/general-approaches-to-insect-control.

Moscardi, F. (1999). Assessment of the application of baculoviruses for the control of Lepidoptera. *Annu. Rev. Entomol.* 44, 257–289.

Murphy, F.A., Faquet, C.M., Bishop, D.H.L., Gabrial, S.A.,Jarvis, A. W. and Martelli, G. P. (1995). Classificationand Nomenclature of Viruses: Sixth Report of the International Committee on Taxonomy of Viruses. Springer-Verlag, Berlin.

Natarajan, K. and Sundaramurthy, V. T. (1990). Effect of neem oil on cotton white fly. Indian *J. Agric. Sci.* 60(4): 290 – 291.

Pigott, C. R and Ellar, D. J. (2007). Role of receptors in *Bacillus thuringiensis* crystal toxin activity.- *Microbiology and Molecular Biology Reviews*, 71: 255-281.

Ramarethinam, S. (2002). Biopesticides – An Overview. In: Resource management in Plant Protection. Vol. I. Published by Plant protection Association of India. NPPTI Campus, Rajendranagar, Hyderabad. pp. 167-179.

Steinhaus, D. C. (1958). Stress as a factor in insect disease. *Proceedings of 10th International Congress of Entomology*, 4: 725 - 730.

Steinhaus, E. A. (1975). Disease in a minor chord. Ohio State University Press, Columbus, USA.

WHO. (2002). Pesticides and Risk of Parkinson Disease: Population Based Cases Control Study. *Arch Neurol.*, 62: 91-95.

Wraight, S. P. and Ramos, M. E. (2015). Delayed efficacy of *Beauveria bassiana* foliar spray applications against Coloradopotato beetle: Impacts of number and timing of applications on larval and next-generation adult populations. *Biological Control*. 2015; 83: 51-67.